Springer Tracts in Modern Physics 101

Springer Tracts in Modern Physics

74 **Solid-State Physics** With contributions by G. Bauer, G. Borstel, H. J. Falge, A. Otto

75 **Light Scattering by Phonon-Polaritons** By R. Claus, L. Merten, J. Brandmüller

76 **Irreversible Properties of Type II Superconductors** By H. Ullmaier

77 **Surface Physics** With contributions by K. Müller, P. Wißmann

78 **Solid-State Physics** With contributions by R. Dornhaus, G. Nimtz, W. Richter

79 **Elementary Particle Physics** With contributions by E. Paul, H. Rollnick, P. Stichel

80* **Neutron Physics** With contributions by L. Koester, A. Steyerl

81 **Point Defects in Metals I:** Introductions to the Theory (2nd Printing)
By G. Leibfried, N. Breuer

82 **Electronic Structure of Noble Metals, and Polariton-Mediated Light Scattering**
With contributions by B. Bendow, B. Lengeler

83 **Electroproduction at Low Energy and Hadron Form Factors**
By E. Amaldi, S. P. Fubini, G. Furlan

84 **Collective Ion Acceleration** With contributions by C. L. Olson, U. Schumacher

85 **Solid Surface Physics** With contributions by J. Hölzl, F. K. Schulte, H. Wagner

86 **Electron-Positron Interactions** By B. H. Wiik, G. Wolf

87 **Point Defects in Metals II:** Dynamical Properties and Diffusion Controlled Reactions
With contributions by P. H. Dederichs, K. Schroeder, R. Zeller

88 **Excitation of Plasmons and Interband Transitions by Electrons** By H. Raether

89 Giant Resonance Phenomena in **Intermediate-Energy Nuclear Reactions**
By F. Cannata, H. Überall

90* **Jets of Hadrons** By W. Hofmann

91 **Structural Studies of Surfaces**
With contributions by K. Heinz, K. Müller, T. Engel, and K. H. Rieder

92 **Single-Particle Rotations in Molecular Crystals** By W. Press

93 **Coherent Inelastic Neutron Scattering in Lattice Dynamics** By B. Dorner

94 **Exciton Dynamics in Molecular Crystals and Aggregates** With contributions by
V. M. Kenkre and P. Reineker

95 **Projection Operator Techniques** in Nonequilibrium Statistical Mechanics
By H. Grabert

96 **Hyperfine Structure in 4d- and 5d-Shell Atoms** By S. Büttgenbach

97 **Elements of Flow and Diffusion Processes in Separation Nozzles** By W. Ehrfeld

98 **Narrow-Gap Semiconductors** With contributions by R. Dornhaus, G. Nimtz, and
B. Schlicht

99 **Dynamical Properties of IV–VI Compounds** With contributions by H. Bilz, A. Bussmann-Holder, W. Jantsch, and P. Vogl

100* **Quarks and Nuclear Forces** Edited by D. C. Fries and B. Zeitnitz

101 **Neutron Scattering and Muon Spin Rotation** With contributions by R. E. Lechner,
D. Richter, and C. Riekel

---

* denotes a volume which contains a Classified Index starting from Volume 36.

Neutron Scattering and Muon Spin Rotation

Contributions by

R. E. Lechner D. Richter C. Riekel

With 118 Figures

Springer-Verlag
Berlin Heidelberg New York Tokyo 1983

Dr. Ruep E. Lechner
Hahn-Meitner-Institut für Kernforschung
Bereich Kernchemie und Reaktor, Glienicker Straße 100
D-1000 Berlin 39, Fed. Rep. of Germany

Dr. Dieter Richter
Institut für Festkörperforschung
Kernforschungsanlage Jülich, Institut 7, Postfach 1913
D-5170 Jülich, Fed. Rep. of Germany

Dr. Christian Riekel
Max-Planck-Institut für Festkörperforschung, Heisenbergstraße 1
D-7000 Stuttgart 80, Fed. Rep. of Germany
Present address: Universität Hamburg, Institut für Anorganische und Angewandte Chemie
Abteilung Angewandte Chemie, Martin-Luther-King-Platz 6
D-2000 Hamburg 13, Fed. Rep. of Germany

Manuscripts for publication should be addressed to:
Gerhard Höhler
Institut für Theoretische Kernphysik der Universität Karlsruhe
Postfach 6380, D-7500 Karlsruhe 1, Fed. Rep. of Germany

Proofs and all correspondence concerning papers in the process of publication should be addressed to:
Ernst A. Niekisch
Haubourdinstrasse 6, D-5170 Jülich 1, Fed. Rep. of Germany

ISBN 3-540-12458-6 Springer-Verlag Berlin Heidelberg New York Tokyo
ISBN 0-387-12458-6 Springer-Verlag New York Heidelberg Berlin Tokyo

Library of Congress Cataloging in Publication Data. Lechner, R. E. (Ruep E.), 1938– . Neutron scattering and muon spin rotation. (Springer tracts in modern physics; 101) 1. Molecular bonds. 2. Neutrons-Scattering. 3. Muons-Spectra. I. Riekel, C. (Christian), 1943– . II. Richter, D. (Dieter), 1947– . III. Title. IV. Series. QCI.S797 vol. 101 [QD461] 539s [541.2'24] 83-14994

Offset printing and bookbinding: Brühlsche Universitätsdruckerei, Giessen
2153/3130 – 5 4 3 2 1 0

Preface

This volume contains two different contributions concerning the application of two of the most advanced experimental techniques in condensed-matter research. Most interesting results from such investigations are presented.

The first contribution (by R.E. Lechner and C. Riekel) treats the application of neutron scattering in condensed-matter research, and particularly to problems in chemistry. Although it is primarily addressed to chemists, it should also be relevant to scientists in physics, metallurgy and biochemistry, since especially in neutron-scattering work the interdisciplinary collaboration between physicists, chemists and others is becoming increasingly important, and it is often difficult to draw clear lines between some of these fields.

Chemistry aims at the preparation and characterization of new materials. In the past, neutron diffraction has become an important analytical tool of the chemist. The rapid evolution of neutron-scattering techniques and of their application to new areas in chemistry, especially in the realm of spectroscopy, calls for a review of latest results from neutron-scattering experiments. With this contribution the authors hope to stimulate new investigations, the planning and execution of which require an understanding of the neutron-scattering technique as much as the direct access to the intellectual resources of the chemist. Rather than aiming at completeness, a number of selected experiments is discussed. Before doing so it is necessary to provide an introduction to the theory of neutron scattering, as it is required for analyzing these experiments. The theory is presented from the point of view of the user, who is more interested in the practical aspect of application than in deductions and proofs. Thus the emphasis is put on the way in which the microscopic properties of systems under study, which are usually represented by suitable models, appear in the theoretical scattering cross-sections. It is hoped that this will enable the reader to understand the original literature describing the experiments, without necessarily having to go through complete books on neutron-scattering theory.

The experiments are discussed in detail with reference to the theoretical treatments and to the original literature. They illustrate new directions of research and recent progress in fields such as chemical bonding and spin densities, conformation of macromolecules, rotational diffusion and tunneling in molecular crystals, diffusion in hydrogen-metal systems and superionic conductors, dynamics of chemical equi-, libria, kinetics of structural transformations, structure and dynamics of physisorbed and chemisorbed adsorbates, intercalation compounds and polymer solutions.

The second contribution (by D. Richter) is concerned with transport mechanisms of light interstitials in metals. Besides the application of neutron scattering to this problem, which is treated comprehensively, results from muon spin rotation (μSR) experiments are discussed extensively.

The investigation of transport properties is one of the most prominent fields in solid state research. Out of the large variety of phenomena this contribution considers the intermediate range between electron-like band motion and classical diffusion which is covered by the diffusional properties of the light interstitials: muon, proton, deuteron and triton in metals. The starting point is an outline of the concept of small polaron motion. Thereby, new developments as small polaron hopping in disordered materials and the influence of phonon fluctuations on the tunneling matrix element between adjacent sites are treated with special emphasis.

On the experimental side muon diffusion experiments have brought about a richness of new and often unexpected results on polaron motion under extreme conditions like very-low temperatures or ultra-high purity of the host. In an exemplary way these results are surveyed and compared with small polaron theory. Other than the outcome of (H, D, T) diffusion experiments, which can be understood in terms of the small polaron concept, new ideas appear to be necessary in order to understand quantum diffusion as performed by the muon.

Neutron scattering is the most important method for a microscopic investigation of H motion. Results on the space and time development of H motion in fcc and bcc metals are presented. Atomistic details of proton trapping are obtained from quasi-elastic and inelastic experiments. Diffusion mechanisms in complex many-component H-storage materials were unraveled. Finally, recent investigations on the dynamics of protons trapped at substitutional and interstitial impurities in Nb are surveyed. These experiments allowed a large extension of the temperature range accessible for H diffusion experiments.

Berlin, Hamburg,
Jülich, March 1983

R.E. Lechner and C. Riekel
D. Richter

Contents

Applications of Neutron Scattering in Chemistry

By R.E. Lechner and C. Riekel (With 49 Figures)

Transport Mechanisms of Light Interstitials in Metals

By D. Richter (With 69 Figures)

Applications of Neutron Scattering in Chemistry*

By R. E. Lechner and C. Riekel

1. Introduction

During the last 30 years neutron scattering has developed into a particularly versatile tool in the investigation of condensed matter. It is used for problems varying from pure physics and chemistry through material science to biochemistry and biology. This development is due partly to the particle wave nature of the neutron, and partly to the specific properties which distinguish neutron radiation from other particle waves (e.g. electrons). The mass of neutrons, $m_n = 1.675 \cdot 10^{-24}$ g, is just big enough to permit those neutrons thermalised in the moderator of a nuclear reactor, and having "thermal" energies of the same order of magnitude as the excitation energies of condensed matter to leave it with wavelengths the same order of magnitude as the interatomic distances in condensed matter. Neutrons are therefore suitable for studying both the static and dynamic structure of condensed matter in microscopic ranges of space (10^{-10}-10^{-5} cm) and time (10^{-14}-10^{-8} s). They thus make an ideal complement to the experimental techniques based on electromagnetic waves which in the same range of space and time, depending on wavelength, have so far been limited either to the investigation of the three-dimensional arrangement of atoms (X-rays, gamma- and short wavelength synchrotron radiation)[1], or in spectroscopy (optical spectroscopy: infrared absorption and Raman scattering) to the study of time-dependent phenomena. The recent development of neutron spectrometers of high energy resolution, in particular at the Institut Laue-Langevin high flux reactor at Grenoble, has linked this technique to the time scale of slower atomic and molecular

*Translated by D.R. Gray, Institut Laue-Langevin, Grenoble, France from the original German edition:
Anwendungen der Neutronenstreuung in der Chemie

[1] The possibility of developing inelastic scattering techniques using synchrotron radiation at high energy transfers ($\hbar\omega \gtrsim 50$ meV) with energy resolutions ($\Delta\hbar\omega$) in the range of 1-10 meV is presently under investigation /1.0/; note that for neutrons actually $10^{-4} \lesssim \Delta\hbar\omega \lesssim 10$ meV.

movements (with characteristic times $\geq 10^{-8}$ s), which can be studied for example by acoustic methods and by nuclear magnetic resonance (NMR).

In addition to the favourably situated measuring range of neutron scattering, there are various advantages resulting from the nature of the interaction of the neutron with matter. The electrical neutrality of the neutron permits considerable penetration depths (in contrast to charged-particle waves such as electrons and ions) making it most suitable in the investigation of volume effects. However, surface effects can also be studied if samples with a high specific area are available. The particular dependence of a neutron's interaction with matter on the type of atomic nucleus generally makes it possible to localise light and heavy atoms with comparable precision. In addition, the spin and the magnetic moment of the neutron enable the study of magnetic phenomena.

A number of books /1.1-7/ and conferences /1.8,9/ have been devoted to the possible applications of neutron scattering, including applications to chemistry. The rapid development of neutron research particularly in this direction calls for a survey of the latest results, which should be of especial interest to chemists. In this volume we make no claim to completeness of coverage, preferring to restrict ourselves to the detailed discussion of a number of selected experiments. We thus hope to encourage further investigations including novel work, whose planning and implementation necessitate both understanding the methods and having access to the chemists' "problems". To introduce this discussion it seems appropriate to explain briefly the scattering experiment, the specific neutron properties mentioned above, the basic scattering theory and the most important dynamic models. This is done in Chaps.2 to 5, followed by a description of the experiments, particularly emphasising molecular and liquid crystals, hydrogen-metal systems, superionic conductors, polymers, intercalation compounds and physisorbed adsorbates.

2. Principle of the Scattering Experiment

When neutron radiation passes through matter, every neutron encountering a nucleus is either absorbed or scattered by it. When scattered it normally continues on its way in a new direction with a different speed and spin orientation, in other words in a different state[2]. Since in condensed matter the scattering nucleus is coupled

[2] After the first scattering process, further collisions with other nuclei may in principle occur. However, by an appropriate choice of sample thickness the probability of this can be kept so low that a correction for multiple scattering is possible. We therefore restrict our discussion to single scattering processes.

to an ensemble of atoms by chemical bonding forces, and since the scattering of the neutron is subject to the conservation laws of energy, momentum and angular momentum, the changes of state of a number of scattered neutrons necessarily form together some kind of mirror image of the possible states of the scattering system. It becomes clear that by measuring these changes of state in a neutron scattering experiment one can study properties of matter concealed in the three-dimensional arrangement of the atoms and their magnetic moments and in their types of bonding. These measurements can thus further our understanding of the relationships between macroscopic and microscopic phenomena.

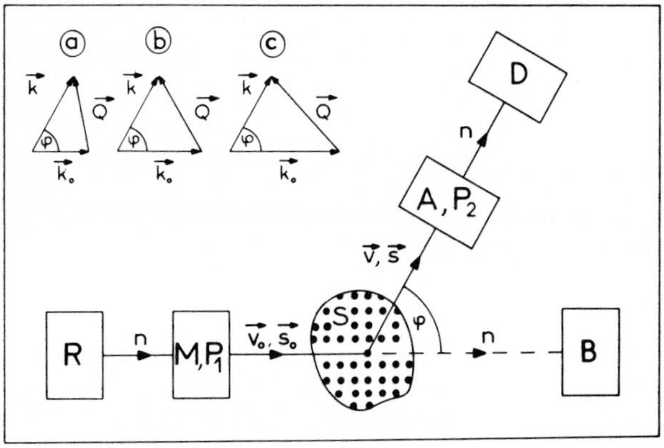

Fig.2.1. Above left: scattering vector diagrams for (a) neutron energy gain, (b) no energy change, (c) neutron energy loss; below right: principle of the scattering experiment with R = neutron source, n = neutrons, M = monochromator, $P_{1,2}$ = polariser, S = sample, B = beam-stop, A = analyser, D = detectors, φ = scattering angle, $\underline{v}_0$, $\underline{s}_0$, $\underline{v}$, $\underline{s}$ ... cf. Table 2.1

Each scattering experiment consists essentially of three parts (Fig.2.1):

a) A beam of neutrons is produced in a well-defined initial state. For this purpose those neutrons which have a given direction of flight within chosen limits are initially selected from the continuous neutron spectrum of the neutron source (R) (e.g. a nuclear reactor). Subsequently – depending on the type of experiment – the desired neutron velocity v_0 is set by means of a monochromator (M) and/or a particular spin orientation s_0 is set by means of a polariser (P_1).

b) The neutron beam prepared in this way falls on the sample S to be investigated. Some of the incident neutrons are scattered by the sample. In addition to controlling the sample environment by varying external parameters (temperature, pressure, magnetic field, etc.), in some cases the scattering power of the sample atoms can be varied by isotope exchange thus increasing the amount of information obtained from the experiment, often without any essential change in the structure of the sample.

c) The intensity of the scattered neutrons is generally measured using detectors (D) as a function of the direction (scattering angle φ) and, if an analyser (A) or a polariser (P_2) is used, also as a function of the velocity v or the spin s. By comparison with the initial state of the neutrons, the statistical probability distribution of the changes in state is then obtained.

In this volume we shall emphasise the two special cases which have so far been used most frequently and with the best success in the study of chemical problems. The first is the inelastic scattering of unpolarised neutrons, where the scattered neutrons are analysed by direction and velocity (Fig.2.1 without polarisers). The equipment used for this can be roughly divided into two classes: triple-axis spectrometers, for which the neutron velocity is selected and measured with the aid of reflection on single crystals; and time-of-flight spectrometers, where the time of flight of the neutrons is used for this purpose. The second special case, neutron diffraction, is obtained from the first, if we do not use velocity analysis[3] This technique is in many respects the precise analogy of X-ray diffraction. Here we shall consider the neutron spectrometer used in each case as a black box with a known function. Interested readers are referred to the appropriate literature /1.2, 3,5;2.1-3/.

Chapter 5 summarises how the desired information on the properties of the sample substance can be derived from the intensity distribution measured (i.e. from the probability distribution of the changes in neutron state). Here we anticipate the definition of the variables usual in the theoretical description of the scattering experiment. We wish to change over to the variables of the neutron energy E and the neutron momentum $\underline{p}$ or the wave vector $\underline{k}$, as these make it possible to present the measured results independently of the particular experiment. The relations between these quantities and the velocity of the neutrons are shown with the usual units in Table 2.1.

Table 2.1. Relationship between neutron velocity and E, $\underline{p}$ and $\underline{k}$

Quantity	Incident neutrons	Scattered neutrons	Usual units
Velocity	$\underline{v}_0$	$\underline{v}$	
Momentum	$\underline{p}_0 = \hbar\underline{k}_0$	$\underline{p} = \hbar\underline{k}$	
Wave vector	$\underline{k}_0$	$\underline{k}$	
Wave number	$k_0 = \lvert\underline{k}_0\rvert = 2\pi/\lambda_0$	$k = \lvert\underline{k}\rvert = 2\pi/\lambda$	$1\ \text{Å}^{-1} = 10^{10}\ \text{m}^{-1}$
Wavelength	λ_0	λ	$1\ \text{Å} = 10^{-1}\ \text{nm}$
Energy	$E_0 = \hbar^2k_0^2/2m$	$E = \hbar^2k^2/2m$	$1\ \text{meV} = 8.07\ \text{cm}^{-1} = 11.6\ \text{K}$

m = neutron mass; $\hbar = h/2\pi$, where h = Planck's constant.

[3] For simplicity we ignore time-of-flight diffraction, whereby velocities are analysed but for which monochromatisation of the incident neutrons is unnecessary.

The change in state of the neutron is generally described by the change $\hbar\omega$ in its energy, the change $\hbar\underline{Q}$ in its momentum and a possible change in spin:

$$\hbar\omega = E - E_0 \quad \text{and} \tag{2.1}$$

$$\hbar\underline{Q} = \hbar(\underline{k} - \underline{k}_0) \quad . \tag{2.2}$$

The relation between scattering vector $\underline{Q}$ and scattering angle φ

$$Q^2 = k^2 + k_0^2 - 2kk_0\cos\varphi \tag{2.3}$$

can be seen from the momentum triangles shown in Fig.2.1. The following three important cases may be distinguished:

a) "Inelastic" scattering with neutron energy gain, where energy is transferred from the sample substance to the neutron: $E > E_0$, $|\underline{k}| > |\underline{k}_0|$.
b) "Elastic" scattering, where no energy is transferred: $E = E_0$, $|\underline{k}| = |\underline{k}_0|$.
c) "Inelastic" scattering with neutron energy loss, where energy is transferred from the neutron to the sample substance: $E < E_0$, $|\underline{k}| < |\underline{k}_0|$.

3. Scattering Cross-Sections

The intensity of the scattered neutrons is determined by the cross-sections of the scattering nuclei for the relevant scattering processes. The total scattering cross-section $\sigma = 4\pi b^2$ is defined as the ratio of the number of neutrons scattered by the atomic nucleus per unit time into the total solid angle 4π over the number of incident neutrons per unit time and unit area. It thus has the dimension of a surface and is quoted in units of 1 barn = 10^{-24} cm^2. The quantity b is called scattering amplitude and has the dimension of a length (scattering length). This is analogous to the atomic form factor f_x, familiar from X-ray scattering, which is proportional to the atomic number Z and is determined by the electronic charge density distribution of the atom. Since the electron shell has spatial dimensions of the order of magnitude of X-ray wavelengths, f_x is a Q-dependent function. The nuclear form factor b occurring in the scattering of thermal neutrons by the atomic nucleus, on the other hand, varies so slowly with Q that it can be regarded as constant in these experiments. This is due to the small dimensions of the atomic nucleus ($\sim 10^{-12}$ cm) relative to the neutron wavelengths used ($\sim 10^{-7}$-10^{-8} cm). A qualitative relationship between b and the spatial extent of the atomic nucleus can be found if the scattering process at the individual atomic nucleus is considered as the consequence of the elastic collision of a neutron with a spherically symmetrical step potential (i.e. with an impenetrable sphere) of radius b. The exact quantum-mechanical calculation for this gives the total cross-section $\sigma = 4\pi b^2$ already mentioned (whereas classical mechanics, here inapplicable, would give only $\sigma = \pi b^2$ for a pointlike neutron).

As there are various types of scattering processes, it is desirable to divide the total scattering cross-section σ into separate parts, whereby we break it down into a "coherent" and an "incoherent" part: $\sigma = \sigma_{coh} + \sigma_{inc}$. The coherent part $\sigma_{coh} = 4\pi b^2_{coh}$ is mainly responsible for the occurrence of the Bragg reflections, which — as with X-ray diffraction — can be used to determine crystal structures.

In these as in other coherent scattering processes the neutron waves scattered at different atoms of the same crystal are capable of interference. The scattered amplitudes are superimposed to give the familiar interference pattern, so enabling the regular arrangement of atoms in the crystal lattice to be determined.

A maximum capacity of interference is present for example in the simplest case when the scattering amplitudes corresponding to the different scattering centres and the phase shifts of the scattered waves caused by the scattering are of uniform size everywhere. If at least one of the two conditions is not met, and if the fluctuations of the scattering amplitudes and/or of the phase shifts about their mean values are randomly distributed over the various scattering centres, only a part of the scattered waves interfere. This (coherent) scattering leads to a diffraction pattern identical to that of a crystal of the same form and with the same lattice, whose lattice positions are filled with (virtual) atoms of the mean scattering amplitude. The remaining scattering, attributable to the fluctuations mentioned above, is designated as incoherent in neutron terminology. In the case of thermal neutrons it is largely independent of the scattering angle φ because of the weak Q dependence of the scattering amplitudes b already mentioned.

To exemplify this phenomenon let us consider the X-ray scattering from the crystal lattice of a solid binary solution AB without short-range order, where the atoms A (scattering amplitude f_A, concentration C_A) and B (scattering amplitude f_B, concentration C_B) are randomly distributed over the lattice positions. Bragg reflection occurs according to the square of the mean scattering amplitude

$$(C_A f_A + C_B f_B)^2 = \langle f \rangle^2 = \sigma^X_{coh}/4\pi \quad , \tag{3.1}$$

while the incoherent scattering (which has nothing to do with the Compton scattering which is also "incoherent") is obtained by subtraction of the coherent from the total scattering:

$$(C_A f_A^2 + C_B f_B^2) - (C_A f_A + C_B f_B)^2 = \langle f^2 \rangle - \langle f \rangle^2 = \sigma^X_{inc}/4\pi \quad . \tag{3.2}$$

We have thus defined the X-ray scattering cross-sections σ^{x}_{coh} and σ^{x}_{inc} analogous to the neutron scattering cross-sections σ_{coh} and σ_{inc}.

Simple rewriting of (3.2) leads to the well-known formula

$$<f^2> - <f>^2 = C_A C_B (f_A - f_B)^2 \tag{3.3}$$

describing this simplest case of defect or disorder scattering known as Laue scattering. It is continuously distributed in between the Bragg reflections. For X-ray scattering, (3.3) has relatively little significance, because ideally disordered solid solutions or mixed crystals are a rarity. However, an analogous expression also applies for incoherent neutron scattering based on the random distribution of the orientations of the nuclear spin and/or the isotopes over the positions of the same type of atom in the crystal. Different isotopes generally have different scattering lengths b_i and these also depend on the orientation of the neutron spin relative to the spin of the scattering nucleus. Generally, isotopes and spin orientations are statistically distributed, as the crystal energy — apart from exceptions at very low temperatures — is independent of these distributions. Then by analogy to definitions (3.1 and 2)

$$\sigma_{coh} = 4\pi b^2_{coh} = 4\pi <b>^2 \quad \text{and} \tag{3.4}$$

$$\sigma_{inc} = 4\pi b^2_{inc} = 4\pi(<b^2> - <b>^2) \quad , \quad \text{where} \tag{3.5}$$

$$<b> = \sum_i c_i b_i \quad , \tag{3.6}$$

$$<b^2> = \sum_i c_i b_i^2 \tag{3.7}$$

and c_i are the concentrations of the isotopes or spin orientations occurring in the crystal. In accordance with this definition $\sigma_{inc} = 0$ when the scattering nuclei have no spin and all scattering lengths are the same.

The incoherent scattering cross-section is thus responsible for that part of the scattered intensity which does not contain any interference effects due to neutron waves scattered by *different* atoms. However, neutron waves scattered by the *same* atom at different times are capable of producing interference. This effect is particularly useful for observing the movement of an individual atom. In many cases both types of scattering produce complementary results and can therefore be advantageously combined.

We now wish to discuss the values of the cross-sections of various atoms, referring to Table 3.1 and Fig.3.1, to emphasise the particular properties of neutron scat-

Table 3.1. Comparison of bound scattering cross-sections and absorption cross-sections in barns (1 barn = 10^{-24} cm^2) for X-rays and neutrons. By analogy with the designations usual for neutrons, $\sigma_{coh} = 4\pi b^2_{coh}$ for the coherent scattering cross-section and σ_{abs} for the absorption cross-section, we define here for X-rays: $\sigma^X_{coh} = 4\pi f^2_X$ and σ^X_{abs}, respectively. Here f_X is the known atomic form factor at maximum value ($\sin\theta = 0$) in the cross-section calculations given in the table

Element	σ^X_{coh} ($\sin\theta = 0$)	σ^X_{abs} (CuKα)	σ_{coh}	σ_{abs} (λ = 1.542 Å)	$\sigma^X_{abs}/\sigma^X_{coh}$	$\sigma_{abs}/\sigma_{coh}$
H	0.985	0.728	1.76	0.285	0.74	0.16
D	0.985	0.728	5.6	~0	0.74	0.0
C	35.9	91.7	5.56	0.003	2.55	$5\cdot10^{-4}$
N	48.8	175	11.1	1.62	3.59	0.15
O	63.6	306	4.23	~0	4.81	0.0
Na	120	1149	1.63	0.434	9.58	0.27
Si	196.1	2826	2.22	0.137	14.41	0.06
Cl	289.5	6293	11.6	28.9	21.55	2.49
Ti	483	16542	1.45	4.99	34.25	3.44
V	530.9	19707	0.031	4.28	37.12	138.06
Ni	784.3	4455	13.3	4.12	5.68	0.31

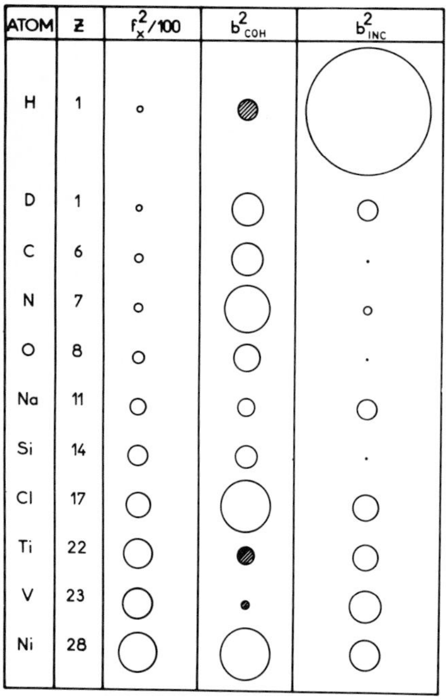

Fig.3.1. Comparison of the bound scattering cross-sections (proportional to the areas of the circles) for 11 selected atoms (atomic charge Z): coherent X-ray scattering cross-section = $4\pi f^2_X$; coherent neutron scattering cross-section = $4\pi b^2_{coh}$; incoherent neutron scattering cross-section = $4\pi b^2_{inc}$. Shaded circles correspond to negative scattering amplitudes (H, Ti and V)

tering in relation to X-ray scattering[4]. Therefore we have selected 11 different atoms whose behavior may be regarded with certain reservations as representative for the whole periodic system. It is immediately striking that the (coherent) scattering cross-sections for X-ray radiation (shown in Table 3.1 and Fig.3.1 for $\sin\theta = 0$) increase with the square of the atomic number. The coherent scattering cross-sections for neutrons, on the other hand, vary about a mean value, and show very little correlation with the atomic number. They are almost all roughly the same order of magnitude. Among the incoherent scattering cross-sections for neutrons, the high value for hydrogen is particularly conspicuous. This is illustrated in Fig. 3.1 by circles whose radii are proportional to the absolute values of the scattering amplitudes, so that the areas of the circles are proportional to the scattering cross-sections. The circles corresponding to the X-ray cross-sections are shown to a scale 1:10 of their size, to permit them to be shown in the same figure as the neutron scattering cross-sections. It should also be emphasised that some atomic nuclei have negative neutron scattering amplitudes (all X-ray scattering amplitudes are positive) and that the neutron scattering amplitudes of different isotopes of the same element are generally different (for X-rays they are of the same size). As an example deuterium is shown with hydrogen in Table 3.1 and Fig. 3.1. Negative signs are indicated in Fig.3.1 by shading the circles. They indicate the phase shift of a scattered wave by π relative to the same wave scattered at the same point on a scattering centre with a positive sign. It is immediately obvious that in a comparison of experimental spectra of samples of the same chemical compound differing only by isotope substitution, the occurrence of different absolute values and/or different signs of the scattering amplitudes for different isotopes can be of great importance in interpreting results. We shall return to this below.

In addition to the values of the scattering cross-sections of different atoms and isotopes in relation to each other, their relationships to the absorption cross-sections of the same atoms are of great importance for the practical performance of scattering experiments. These relationships are shown in Table 3.1 for the elements selected together with the coherent scattering cross-sections and the absorption cross-sections using as an example CuKα X-rays ($\lambda = 1.542$ Å), with the neutron absorption cross-sections for the same wavelength indicated. It can be seen that the X-ray cross-sections σ^X_{coh}, σ^X_{abs} are generally considerably larger than the corre-

[4] The scattering cross-sections are discussed in terms of "bound" cross-sections and "bound" scattering lengths, respectively, which correspond to fixed atoms. The relation between these and the actual total scattering cross-sections of atoms (which generally are not rigidly bound in real systems) can be obtained by integration of the differential scattering cross-sections treated in Chap.4. Detailed discussions of the cross-sections of all atoms and isotopes as far as they are known may be found in /2.2;3.1/.

sponding values for neutrons, σ_{coh} and σ_{abs}. In addition, however, the ratio $\sigma^x_{abs}/\sigma^x_{coh}$ is, with certain exceptions, greater or much greater than 1, whereas for neutrons the opposite is the case for many atoms. The result is that the (usually undesired) effect of absorption is generally much more important in X-ray scattering experiments than with neutron scattering.

4. Scattering Theory

4.1 Differential Cross-Sections and Scattering Functions

Following the above discussion of the integral scattering cross-sections σ_{inc} and σ_{coh} we can now concentrate on the quantities which are the direct aim of the scattering experiment, i.e. the differential scattering cross-sections. A distinction is made by number of variables, as a function of which the scattered intensity is analysed, between differential and partial differential cross-sections. The general relationships between these quantities and the structure and dynamics of condensed matter are discussed in detail in the literature /1.5;2.1,2;4.1-4/. Here we must limit this discussion to a few major points. Our main attention will be directed to the pertinent general formulae used to interpret the experimentally observed intensities with the aid of suitable special models of the static and dynamic structure of condensed matter. We begin with the double differential cross-section for inelastic scattering of unpolarised neutrons.

The quantum-mechanical form of this scattering cross-section (per atom) for a scattering system consisting of N atoms is as follows:

$$\frac{\partial^2\sigma}{\partial\Omega\partial(\hbar\omega)} = \frac{1}{\hbar}\cdot\frac{k}{k_0}\cdot\frac{1}{N}\sum_{i,j}^{N}\sum_{\psi_0}p(\psi_0)\sum_{\psi_1}\langle\psi_0|b_j e^{-i\underline{Q}\underline{r}_j}|\psi_1\rangle\langle\psi_1|b_i e^{i\underline{Q}\underline{r}_i}|\psi_0\rangle$$

$$\cdot\,\delta[(E_{\psi1}-E_{\psi0})/\hbar-\omega] \quad . \tag{4.1}$$

Here we average over all pairs (i,j) of atoms with the bound scattering lengths (b_i,b_j) and the positions $(\underline{r}_i,\underline{r}_j)$, and sum over all the sample states ($\langle\psi_0|$ before and $|\psi_1\rangle$ after the scattering process) characterised by the wave functions ψ; $p(\psi_0)$ is the statistical weight of the initial state ψ_0, i.e. the Boltzmann distribution $(1/Z)\exp(-E_{\psi0}/k_BT)$ normalised by the partition function Z. The δ function expresses the conservation of the total energy (of neutron plus scattering system). The wave function ψ and the energy eigenvalues E_ψ are calculated from the Schrödinger equation. When this has been done, the major part of the work is completed and the double differential scattering cross section can be determined according to (4.1).

If polarised neutrons are used, or the spins of the atomic nuclei are not statistically distributed, or there is a correlation between the spin states of the scattering system before and after the scattering process, the wave functions ψ in (4.1) must also describe the possible spin states of the scattering system. In addition, the scattering lengths b_i then include terms depending both on the neutron spin and on the nuclear spin; the right-hand side of (4.1) must then be averaged over the initial and final spin states of the neutron.

If the system under consideration consists of only one type of atom and if the isotopes and nuclear spin orientations are statistically distributed, there is no correlation between the values of the scattering lengths b_j and the atomic positions $\underline{r}_j$. The averages over isotope and spin orientation distributions and over the initial states of the system can then be calculated independently of each other, and (4.1) simplifies to:

$$\frac{\partial^2\sigma}{\partial\Omega\partial(\hbar\omega)} = \frac{1}{\hbar}\cdot\frac{k}{k_0}\cdot\frac{1}{N}\cdot\sum_{i,j}\langle b_j b_i\rangle\sum_{\psi_0}p(\psi_0)$$

$$\cdot\sum_{\psi_1}\langle\psi_0|e^{-i\underline{Q}\underline{r}_j}|\psi_1\rangle\langle\psi_1|e^{i\underline{Q}\underline{r}_i}|\psi_0\rangle\cdot\delta[(E_{\psi1}-E_{\psi0})/\hbar-\omega] \quad \text{with} \qquad (4.2)$$

$$\langle b_j\cdot b_i\rangle = \langle b\rangle^2 \text{ for } j \neq i \qquad \text{and}$$

$$\langle b_j\cdot b_i\rangle = \langle b^2\rangle \text{ for } j = i \quad ,$$

so that in general

$$\langle b_j\cdot b_i\rangle = \langle b\rangle^2 + \delta_{ij}(\langle b^2\rangle - \langle b\rangle^2) = b^2_{coh} + \delta_{ij}b^2_{inc} \qquad (4.3)$$

(δ_{ij} = Kronecker symbol).

By using (4.3) the scattering cross-section (4.2) can be separated into a coher- and an incoherent part:

$$\frac{\partial^2\sigma}{\partial\Omega\partial(\hbar\omega)} = \frac{1}{\hbar}\cdot\frac{k}{k_0}\left[b^2_{coh}\cdot S_{coh}(\underline{Q},\omega) + b^2_{inc}\cdot S_{inc}(\underline{Q},\omega)\right] \quad . \qquad (4.4)$$

This separation has the following physical significance: the different atomic nuclei of the scattering system have generally different scattering lengths. The scattered intensity is now considered to consist of two contributions. The first term is the (coherent) scattering, which the same system (with identical atomic positions and movements) would provide if all atomic nuclei had the same averaged scattering length b_{coh}. The second term is the other (incoherent) scattering of the actual system, which comes from the statistical distribution of the deviations of the scattering length from its mean value.

Here $S_{coh}(\underline{Q},\omega)$ and $S_{inc}(\underline{Q},\omega)$ are called coherent and incoherent scattering functions, respectively. Note that these scattering functions alone contain all the information on the interatomic interactions in the sample. They depend neither on the mass and energy of the neutrons nor on interaction of the neutrons with the scattering centres. The other coefficients, on the other hand, characterise the incident (k_0) and scattered (k) particles and the neutron-nucleus interaction (b_{coh}, b_{inc}). In other words, the scattering functions are pure material properties, completely independent of any particular experimental technique; e.g. $S_{coh}(\underline{Q},\omega)$ is the experimentally determined quantity also in (coherent) light, X-ray or Mößbauer scattering, when the scattered radiation is analysed with respect to energy, and $S_{inc}(\underline{Q},\omega)$ can also be observed with Mößbauer absorption.

The simple factorisation of material properties and neutron-nucleus interaction according to (4.4) is however possible only when the scattering lengths b_j are statistically distributed over the atomic positions $\underline{r}_j$. A generalisation can be made without major difficulties to polyatomic systems and/or non-statistical distributions of scattering lengths [cf. (4.1)]. In the theoretical discussion, however, we limit ourselves to the simplest case.

Explicitly the coherent scattering function reads

$$S_{coh}(\underline{Q},\omega) = \frac{1}{N}\sum_{i,j}^{N}\sum_{\psi_0} p(\psi_0)\sum_{\psi_1}\langle\psi_0|e^{-i\underline{Q}\underline{r}_j}|\psi_1\rangle \cdot \langle\psi_1|e^{i\underline{Q}\underline{r}_i}|\psi_0\rangle \cdot \delta[(E_{\psi 1}-E_{\psi 0})/\hbar-\omega] \quad . \tag{4.5}$$

To obtain the incoherent scattering function, it is sufficient to replace $\sum_{i,j}$ by $\sum_i$ and $\underline{r}_j$ by $\underline{r}_i$ in (4.5):

$$S_{inc}(\underline{Q},\omega) = \frac{1}{N}\sum_{i=1}^{N}\sum_{\psi_0} p(\psi_0)\sum_{\psi_1}|\langle\psi_1|e^{i\underline{Q}\underline{r}_i}|\psi_0\rangle|^2 \cdot \delta[(E_{\psi 1}-E_{\psi 0})/\hbar-\omega] \quad . \tag{4.6}$$

The quantum-mechanical expressions (4.5) and (4.6) are of practical importance particularly when — at a fixed value of Q — the scattering function contains a finite number of sharp (ideally delta-function-like) maxima, which may be assigned to well-defined transitions between discrete energy levels $E_{\psi 0}$ and $E_{\psi 1}$ of the scattering system. It is of course a necessary condition that the eigenstates of the unperturbed sample, characterised by ψ_0 and ψ_1, are known or can be calculated. This applies for example to harmonic oscillators, harmonic crystals (phonons) and magnetic systems (spin waves) as well as to rotational tunneling /4.5,6/ which is observed at low temperatures in molecular crystals (Sects.6.6 and 7.4.2).

If the quantum-mechanical nature of the scattering function is largely negligible (in general at higher temperatures and not too large values of $\hbar\omega$ and Q), the dynamics of the system under investigation can be treated classically, so often facilitating calculation of the scattering cross-section, but a quantum-mechanical correction factor is generally necessary. For this purpose it is useful to change from the eigenfunction formulation used so far (4.1,2,5,6) to another of equally general validity, in which the emphasis is on the space- and time-dependent correlation of the probability density distributions of the atoms. The correlation functions thus occurring were introduced by VAN HOVE /4.7/.

4.2 Van Hove Correlation Functions

Van Hove's theory starts from the generalisation of the space-dependent pair distribution function $g(\underline{r})$ known in X-ray diffraction which describes the average probability density distribution of the scattering system as seen from the position of a particle. Under certain conditions, fulfilled in our context, the double-differential scattering cross-section can be formulated with the aid of the space- and time-dependent pair distribution function $G(\underline{r},t)$ (pair correlation function) and the function $G_s(\underline{r},t)$ (auto-correlation or self-correlation function).

Similar to the relation between the differential scattering cross-section $d\sigma/d\Omega$ of X-ray diffraction and $g(\underline{r})$ via a space Fourier transform, $\partial^2\sigma/\partial\Omega\partial(\hbar\omega)$ [cf. (4.4)] can be defined with the aid of the space-time Fourier transformations of $G(\underline{r},t)$ and $G_s(\underline{r},t)$. The following expressions are thus obtained for the scattering functions:

$$S(\underline{Q},\omega) = \frac{1}{2\pi} \int_{-\infty}^{\infty} e^{-i\omega t} \int e^{i\underline{Q}\underline{r}}\, G(\underline{r},t)d\underline{r}dt \quad , \tag{4.7}$$

$$S_s(\underline{Q},\omega) = \frac{1}{2\pi} \int_{-\infty}^{\infty} e^{-i\omega t} \int e^{i\underline{Q}\underline{r}}\, G_s(\underline{r},t)d\underline{r}dt \quad . \tag{4.8}$$

In the reverse direction it is possible to obtain the Van Hove correlation functions from the scattering functions by inversion of these Fourier transforms:

$$G(\underline{r},t) = \frac{1}{(2\pi)^3} \int_{-\infty}^{\infty} e^{i\omega t} \int e^{-i\underline{Q}\underline{r}} S(\underline{Q},\omega)d\underline{Q}d\omega \quad , \tag{4.9}$$

$$G_s(\underline{r},t) = \frac{1}{(2\pi)^3} \int_{-\infty}^{\infty} e^{i\omega t} \int e^{-i\underline{Q}\underline{r}} S_s(\underline{Q},\omega)d\underline{Q}d\omega \quad . \tag{4.10}$$

Here, to simplify the notation, instead of $S_{coh}(\underline{Q},\omega)$ and $S_{inc}(\underline{Q},\omega)$ we have used the designations $S(\underline{Q},\omega)$ and $S_s(\underline{Q},\omega)$, which are also used in the literature, and we shall continue to do so throughout this volume.

Using the example of (4.5,7) for the coherent scattering function, we shall now explain the relationship between the two formulations on which these expressions are based. First we replace in (4.5) the δ function standing for energy conservation by its integral representation

$$\delta[(E_{\psi 1}-E_{\psi 0})/\hbar-\omega] = \frac{1}{2\pi}\int_{-\infty}^{\infty} \exp\left[\frac{i}{\hbar}(E_{\psi 1}-E_{\psi 0}-\hbar\omega)t\right]dt$$

$$= \frac{1}{2\pi}\int_{-\infty}^{\infty} e^{-i\omega t} \exp\left[\frac{i}{\hbar}(E_{\psi 1}-E_{\psi 0})t\right]dt \quad . \tag{4.11}$$

Instead of (4.5) we thus obtain

$$S(\underline{Q},\omega) = \frac{1}{2\pi}\int_{-\infty}^{\infty} e^{-i\omega t}\,\frac{1}{N}\sum_{i,j}^{N}\sum_{\psi_0} p(\psi_0)\sum_{\psi_1} \langle\psi_0|e^{-i\underline{Q}\underline{r}_j}|\psi_1\rangle$$

$$\cdot\langle\psi_1|e^{i\underline{Q}\underline{r}_i}|\psi_0\rangle \exp\left[\frac{i}{\hbar}(E_{\psi 1}-E_{\psi 0})t\right]dt \quad . \tag{4.12}$$

If we now use the closure relation of the wave functions, the definitions of the time-dependent Heisenberg operators $\underline{r}_i(t)$ and the thermal average of a quantum-mechanical operator denoted by <operator>, we obtain the coherent scattering function often written as follows (this calculation is given in more detail in Appendix A):

$$S(\underline{Q},\omega) = \frac{1}{2\pi}\int_{-\infty}^{\infty} e^{-i\omega t}\,\frac{1}{N}\sum_{i,j}^{N}\langle\exp[-i\underline{Q}\underline{r}_j(0)]\cdot\exp[i\underline{Q}\underline{r}_i(t)]\rangle\, dt \quad . \tag{4.13}$$

Defining further the "intermediate coherent scattering function"

$$I(\underline{Q},t) = \frac{1}{N}\sum_{i,j}\langle\exp[-i\underline{Q}\underline{r}_j(0)]\,\exp[i\underline{Q}\underline{r}_i(t)]\rangle \quad , \tag{4.14}$$

we obtain

$$S(\underline{Q},\omega) = \frac{1}{2\pi}\int_{-\infty}^{\infty} e^{-i\omega t}\, I(\underline{Q},t)dt \quad . \tag{4.15}$$

Comparison with (4.7) shows that

$$I(\underline{Q},t) = \int e^{i\underline{Q}\underline{r}}\, G(\underline{r},t)d\underline{r} \quad . \tag{4.16}$$

By analogy we can move from the eigenfunction formulation of the incoherent scattering function (4.6) to the Van Hove formulation (4.8) and obtain

$$S_s(\underline{Q},\omega) = \frac{1}{2\pi}\int_{-\infty}^{\infty} e^{-i\omega t}\, I_s(\underline{Q},t)dt \quad , \tag{4.17}$$

with

$$I_s(\underline{Q},t) = \frac{1}{N}\sum_{i}\langle\exp[-i\underline{Q}\underline{r}_i(0)]\,\exp[i\underline{Q}\underline{r}_i(t)]\rangle \quad . \tag{4.18}$$

Comparing (4.8) with (4.17) also shows that

$$I_s(\underline{Q},t) = \int e^{i\underline{Q}\underline{r}}\, G_s(\underline{r},t)d\underline{r} \quad . \tag{4.19}$$

This function is called the intermediate incoherent scattering function. The intermediate scattering functions are thus the spatial Fourier transforms of $G(\underline{r},t)$ and $G_s(\underline{r},t)$. In practice it is sometimes more advantageous to start from $I(\underline{Q},t)$ and $I_s(\underline{Q},t)$ than to calculate the Van Hove correlation functions directly. However, the latter are more easily accessible for physical interpretation.

4.3 Classical Approximation and Interpretation of the Correlation Functions

If (4.15) is inserted into (4.9), we obtain

$$G(\underline{r},t) = \frac{1}{(2\pi)^3} \int e^{-i\underline{Q}\underline{r}} I(\underline{Q},t)d\underline{Q} \tag{4.20}$$

where use has been made of

$$\int_{-\infty}^{\infty} e^{i\omega(t-t')} d\omega = 2\pi\delta(t-t') \quad . \tag{4.21}$$

Using (4.14) we then obtain

$$G(\underline{r},t) = \frac{1}{(2\pi)^3} \frac{1}{N} \sum_{i=1}^{N} \sum_{j=1}^{N} \int e^{-i\underline{Q}\underline{r}} \langle \exp[-i\underline{Q}\underline{r}_j(0)] \cdot \exp[i\underline{Q}\underline{r}_i(t)] \rangle d\underline{Q} \quad . \tag{4.22}$$

Applying the convolution theorem for Fourier transforms to the product in angular brackets < > gives the following result (Appendix B):

$$G(\underline{r},t) = \frac{1}{N} \sum_{i,j}^{N} \int \langle \delta[\underline{r}+\underline{r}_j(0)-\underline{r}'] \cdot \delta[\underline{r}'-\underline{r}_i(t)] \rangle d\underline{r}' \quad . \tag{4.23}$$

This fundamental quantum-mechanical expression for the pair correlation function $G(\underline{r},t)$ contains the Heisenberg operators $\underline{r}_j(0)$ and $\underline{r}_i(t)$ which do not commute for $t \neq 0$; therefore the involved convolution integrals cannot be obtained immediately.

Let us now consider $G(\underline{r},t)$ at the instant $t = 0$. In this special case the operators commute and we can write

$$G(\underline{r},0) = \sum_{i,j}^{N} \langle \delta[\underline{r}+\underline{r}_j(0)-\underline{r}_i(0)] \rangle = \delta(\underline{r})+g(\underline{r}) \quad , \tag{4.24}$$

where

$$g(\underline{r}) = \sum_{i \neq j} \langle \delta(\underline{r}+\underline{r}_j-\underline{r}_i) \rangle \tag{4.25}$$

is the static pair distribution function well-known from X-ray diffraction; $g(\underline{r})d\underline{r}$ is the probability of finding any atom in the volume element $d\underline{r}$ at position $\underline{r}$ if another atom is at the origin[5]. By analogy we obtain

$$G_s(\underline{r},0) = \delta(\underline{r}) \quad . \tag{4.26}$$

If "classical" conditions are present (i.e. if the transferred energy $\hbar\omega$ and the transferred momentum $\hbar\underline{Q}$ are related with the thermal energy per degree of freedom $k_BT/2$ by the inequalities $|\hbar\omega| \ll k_BT/2$ and $(\hbar Q)^2/2M \ll k_BT/2$, respectively, where M is the mass of the atom), the quantum-mechanical nature of the scattering function can be neglected. Then $\underline{r}_j(0)$ and $\underline{r}_i(t)$ are no longer regarded as quantum-mechanical operators, but simply as the position vectors of the atoms j and i at the times t = 0 and t, respectively. It is now possible to integrate over $\underline{r}'$, so that

$$G^{cl}(\underline{r},t) = \frac{1}{N}\sum_{i=1}^{N}\sum_{j=1}^{N}\langle\delta[\underline{r}+\underline{r}_j(0)-\underline{r}_i(t)]\rangle \quad . \tag{4.27}$$

If we assume for simplicity that all atoms are dynamically equivalent, averaging over j becomes trivial and gives only the factor N, so that (with arbitrary j)

$$G^{cl}(\underline{r},t) = \sum_{i=1}^{N}\langle\delta[\underline{r}+\underline{r}_j(0)-\underline{r}_i(t)]\rangle \quad . \tag{4.28}$$

Whereas the thermal average of the operator in (4.23) indicated by <...> covers quantum-mechanical averaging over the final ψ_1 and initial ψ_0 states of the scattering system at temperature T (with statistical weighting according to the Boltzmann distribution), in (4.27,28) we have purely statistical averaging. The interpretation of the classical pair correlation function follows from this, according to which $G^{cl}(\underline{r},t)d\underline{r}$ is the probability of finding any atom at time t in the volume element $d\underline{r}$ at point $\underline{r}$, if a particular atom was at the origin at the instant t = 0. This probability is averaged over all possible points of origin.

Similarly, inserting (4.17) in (4.10) gives first the relationship between the auto-correlation function and the incoherent intermediate scattering function

$$G_s(\underline{r},t) = \frac{1}{(2\pi)^3}\int e^{-i\underline{Q}\underline{r}}\, I_s(\underline{Q},t)d\underline{Q} \tag{4.29}$$

and then with (4.18) and the convolution theorem,

$$G_s(\underline{r},t) = \frac{1}{N}\sum_{i=1}^{N}\int\langle\delta[\underline{r}+\underline{r}_i(0)-\underline{r}']\cdot\delta[\underline{r}'-\underline{r}_i(t)]\rangle\, d\underline{r}' \quad . \tag{4.30}$$

The classical form of this expression for a system of dynamically equivalent atoms (where i is the label of any atom) is

[5] Here $d\underline{r}$ is the symbol usually employed in the literature for the differential of the volume in which the vector $\underline{r}$ is allowed to vary (cf., e.g., /4.4/ p. 12).

$$G_s^{cl}(\underline{r},t) = \langle\delta[\underline{r}+\underline{r}_i(0)-\underline{r}_i(t)]\rangle \quad . \tag{4.31}$$

To write explicitly the statistical averaging <...> to be carried out here, it is useful to introduce the following new definitions

$\underline{r}_0 = \underline{r}_i(0)$ = position of the ith atom at time $t = 0$;
$\underline{r}_i = \underline{r}_i(t)$ = position of the ith atom at time t;
$P(\underline{r}_i-\underline{r}_0,\underline{r}_0,t)d\underline{r}_i$ = conditional probability that the same particle which was at the origin $\underline{r}_0$ at time $t = 0$ is in the volume element $d\underline{r}_i$ *at position* $\underline{r}_i$ at time t;
$p(\underline{r}_i) = \lim_{t\to\infty} P(\underline{r}_i-\underline{r}_0,\underline{r}_0,t)$; this limiting value is independent of $\underline{r}_0$ (the origin $\underline{r}_0$ is "forgotten" for $t\to\infty$) and is identical to the probability density distribution of all possible origins $\underline{r}_0$ of the ith atom, $p(\underline{r}_0)$.

With the aid of these definitions we can write

$$G_s^{cl}(\underline{r},t) = \int d\underline{r}_0 \int d\underline{r}_i P(\underline{r}_i-\underline{r}_0,\underline{r}_0,t)p(\underline{r}_0)\cdot\delta[\underline{r}+\underline{r}_0-\underline{r}_i] \tag{4.32}$$

and finally, after the integration over $\underline{r}_i$ has been carried out,

$$G_s^{cl}(\underline{r},t) = \int d\underline{r}_0 P(\underline{r},\underline{r}_0,t)p(\underline{r}_0) \quad . \tag{4.33}$$

Here $P(\underline{r},\underline{r}_0,t)$ is now the conditional probability density that the atom which at time $t = 0$ was at the origin $\underline{r}_0$ is *at the distance* $\underline{r}$ from the origin at time t. In accordance with this interpretation the classical (but also the quantum-mechanical) correlation functions have the following important integral properties:

$$\int G^{cl}(\underline{r},t)d\underline{r} = N \text{ and } \int G_s^{cl}(\underline{r},t)d\underline{r} = 1 \quad . \tag{4.34}$$

Interpreting $G^{cl}(\underline{r},t)$ and $G_s^{cl}(\underline{r},t)$ as probability density functions permits their calculation from clear physical models of the systems under investigation. The associated scattering functions $S^{cl}(\underline{Q},\omega)$ and $S_s^{cl}(\underline{Q},\omega)$ are obtained by Fourier transformation according to (4.7,8), respectively. In particular, with (4.8) the incoherent scattering function is obtained in the following form which is very useful for practical applications (Chap.5):

$$S_s^{cl}(\underline{Q},\omega) = \frac{1}{2\pi}\int_{-\infty}^{\infty} e^{-i\omega t}dt \int d\underline{r}_0 p(\underline{r}_0)\int e^{i\underline{Q}\underline{r}} P(\underline{r},\underline{r}_0,t)d\underline{r} \quad . \tag{4.35}$$

When the classical approximation is used it must be noted that the scattering functions always behave classically for long times t and for large values of $\underline{r}$ (i.e. for small energies $\hbar\omega$ and momentum transfers $\hbar\underline{Q}$, respectively; cf. the limits mentioned above), whereas on the other hand for short times and small distances (i.e. for large values of $\hbar\omega$ and $\hbar\underline{Q}$) quantum effects are to be expected. As the exact quantum-mechanical calculation often cannot be carried out, the classically calculated scattering function must – if necessary – be corrected for such effects.

The most important correction, which is sufficient for our purposes, results from the relation between the scattering cross-sections in the cases of energy gain ($E > E_0$) and energy loss ($E < E_0$) of the neutron. If two different energy states of a scattering system are considered, it can be shown that the two probabilities for the transitions caused by neutron scattering from the first to the second or from the second to the first state are equal, provided that both states are equally occupied. However, this is generally true only for high temperatures since the relationship of the occupation numbers is given by the Boltzmann distribution. If the energy difference between the two states is $\hbar\omega$, in thermal equilibrium the probability that the system is in the higher energy state is less by a factor $\exp(\hbar\omega/k_BT)$ than the probability that it is in the lower energy state $[\exp(\hbar\omega/k_BT) \to 1$ for $T \to \infty]$. It follows that the scattering functions for energy gain and energy loss processes are always linked by the following detailed balance condition:

$$S(-\underline{Q},-\omega) = \exp(+\hbar\omega/k_BT) \cdot S(\underline{Q},\omega) \quad . \tag{4.36}$$

Consequently $S(\underline{Q},\omega)$ is not symmetrical in ω.

A function symmetrical in ω, like the classical scattering function $S^{cl}(\underline{Q},\omega)$, is obviously obtained if (4.36) is multiplied by the factor $\exp(-\hbar\omega/2k_BT)$. If the function then obtained as the right-hand side of (4.36) is identified with $S^{cl}(\underline{Q},\omega)$, in many cases a very good approximation of the actual scattering function $S(\underline{Q},\omega)$ is obtained:

$$S^{cl}(\underline{Q},\omega) = \exp(+\hbar\omega/2k_BT) \cdot S(\underline{Q},\omega) \quad . \tag{4.37a}$$

We can then write

$$S(\underline{Q},\omega) = \exp(-\hbar\omega/2k_BT) \cdot S^{cl}(\underline{Q},\omega) \quad . \tag{4.37b}$$

Similarly, for the incoherent scattering function

$$S_s(\underline{Q},\omega) = \exp(-\hbar\omega/2k_BT) \cdot S_s^{cl}(\underline{Q},\omega) \quad . \tag{4.38}$$

These are the functions which are generally used to interpret scattering experiments on classical systems. The calculation of the incoherent scattering function for specific (classical) models will be treated in more detail in Sect.5. Designing similar models for the coherent scattering function is of course much more difficult, as this requires detailed assumptions on the time-dependent correlation between different atoms. An approximation avoiding this problem is mentioned in Sect.7.3.

In the diffraction experiment the scattered intensity is analysed not by energy but only by scattering angle. With a system of identical atoms with statistical distribution of isotopes and nuclear spin orientations this corresponds to an integration of (4.4) over $\hbar\omega$:

$$\frac{d\sigma}{d\Omega} = b^2_{coh} \int_{-\infty}^{+\infty} \frac{k}{k_0} S(\underline{Q},\omega)d\omega + b^2_{inc} \int_{-\infty}^{+\infty} \frac{k}{k_0} S_s(\underline{Q},\omega)d\omega \quad . \tag{4.39}$$

If now, as with X-ray diffraction, $k \cong k_0$ for the whole range of relevant energy transfers, it may be assumed that $k = k_0$ (static approximation):

$$\frac{d\sigma}{d\Omega} = \{b^2_{coh} S(\underline{Q}) + b^2_{inc}\} \quad , \tag{4.40}$$

where we used the integral properties of the scattering functions

$$\int_{-\infty}^{\infty} S(\underline{Q},\omega)d\omega = S(\underline{Q}) = I(\underline{Q},0) \tag{4.41}$$

and

$$\int_{-\infty}^{\infty} S_s(\underline{Q},\omega)d\omega = 1 = I_s(\underline{Q},0) \quad . \tag{4.42}$$

From this and using (4.16,24) the following simple relationship between the structure factor $S(\underline{Q})$ and the static pair distribution function $g(\underline{r})$ is obtained:

$$S(\underline{Q}) = 1 + \int g(\underline{r})\exp(i\underline{Q}\underline{r})d\underline{r} \quad . \tag{4.43}$$

The static approximation is also pertinent in neutron scattering, if the energy of the incident neutrons is sufficiently high. However, if this condition is not met, (4.40) does not apply strictly, but a corresponding correction can generally be made. This is important when S(Q) for liquids (liquid structure) is measured by neutrons /4.8/.

In the determination of crystal structures the aim of the experiment is generally not the function $S(\underline{Q})$, but its purely elastic part $S^{el}_{coh}(\underline{Q},\omega)$. For this $k = k_0$ is of course always true also with neutrons, so that the static approximation is not necessary here.

The functions $S(\underline{Q})$, $S^{el}_{coh}(\underline{Q},\omega)$ as well as the purely elastic part $S^{el}_{inc}(\underline{Q},\omega)$ of the incoherent scattering function give an idea of the spatial arrangement of the atoms in the system under investigation through their relationships with the associated probability density distributions. Knowing these functions is therefore an important premise, without which it will generally be difficult to produce a realistic model of the dynamics of the system studied [which is described by means of the complete scattering functions $S(\underline{Q},\omega)$ and $S_s(\underline{Q},\omega)$].

To clarify the relationship between elastic scattering and the Van Hove correlation functions, let us now consider these in the limit of very long times ($t\to\infty$). As the systems to be investigated are always large enough to obey the laws of statistical physics, we may assume that there is no correlation between the positions of particles, which are observed separated from each other by sufficiently large time intervals. For $t\to\infty$ we can therefore replace the mean value of the product of δ functions in (4.23) by a product of average values:

$$G(\underline{r},t\to\infty) = \frac{1}{N}\int d\underline{r}' < \sum_{j=1}^{N} \delta[\underline{r}+\underline{r}_j(0)-\underline{r}'] > < \sum_{i=1}^{N} \delta[\underline{r}'-\underline{r}_i(t)] > \quad . \tag{4.44}$$

If we use the definition of the space- and time-dependent particle density function

$$\rho(\underline{r},t) = \sum_{i=1}^{N} \delta[\underline{r}-\underline{r}_i(t)] \tag{4.45}$$

and consider that this becomes independent of time for $t\to\infty$, we have

$$G(\underline{r},\infty) = \frac{1}{N}\int d\underline{r}' <\rho(\underline{r}')><\rho(\underline{r}-\underline{r}')> \quad . \tag{4.46}$$

This function describes the self-correlation of the particle density. It is an important aid in determining crystal structures and is essentially the well-known Patterson function. The elastic part $S^{el}_{coh}(\underline{Q},\omega)$ of the coherent scattering function can now be expressed by the Fourier transform of $<\rho(\underline{r})>$. If we divide $G(\underline{r},t)$ into an asymptotic time-independent and a time-dependent part, so that

$$G(\underline{r},t) = G(\underline{r},\infty) + G'(\underline{r},t) \quad , \tag{4.47}$$

and insert this into (4.7), we clearly obtain the sum of two terms, the first of which is purely elastic and the second inelastic. The elastic term is

$$S^{el}_{coh}(\underline{Q},\omega) = \delta(\omega)\int e^{i\underline{Q}\underline{r}}\, G(\underline{r},\infty)d\underline{r} \quad , \tag{4.48}$$

$$S^{el}_{coh}(\underline{Q},\omega) = \delta(\omega)\cdot I(\underline{Q},\infty) \qquad \text{with} \tag{4.49}$$

$$I(\underline{Q},\infty) = \int e^{i\underline{Q}\underline{r}}\, \frac{1}{N}\int d\underline{r}' <\rho(\underline{r}')><\rho(\underline{r}-\underline{r}')>\, d\underline{r} \quad . \tag{4.50}$$

With the aid of the substitution $\underline{r}'' = \underline{r} - \underline{r}'$ we obtain

$$I(\underline{Q},\infty) = \frac{1}{N}\int d\underline{r}''\int d\underline{r}' <\rho(\underline{r}')><\rho(\underline{r}'')> e^{i\underline{Q}(\underline{r}'+\underline{r}'')} \tag{4.51}$$

and finally

$$S^{el}_{coh}(\underline{Q},\omega) = \delta(\omega)\,\frac{1}{N}\,\left|\int d\underline{r}\, e^{i\underline{Q}\underline{r}} \langle\rho(\underline{r})\rangle\right|^2 \quad . \tag{4.52}$$

In crystals the time-averaged density distribution $\langle\rho(\underline{r})\rangle$ is of course periodic in space (long-range order) and one can therefore write (4.52) simply as a sum of Bragg reflections if the possibility of static disorder is ignored:

$$S^{el}_{coh}(\underline{Q},\omega) = \delta(\omega)\,\frac{1}{N}\,\frac{(2\pi)^3}{V_E}\sum_{\tau} |F(2\pi\underline{\tau})|^2\,\delta(\underline{Q}-2\pi\underline{\tau}) \quad , \tag{4.53}$$

where V_E is the volume of the unit cell and $\sum_{\tau}$ extends over all reciprocal lattice vectors $\underline{\tau}$. As is well known, every vector $2\pi\underline{\tau}$ ends in a reciprocal lattice point, which is identified by Miller's indices (hkl) /2.2/.

The structure factor $F(2\pi\underline{\tau})$ contains information on the time-averaged atomic density distribution in the unit cell relevant in determining the crystal structure.

The experimental determination of $S^{el}_{coh}(\underline{Q},\omega)$ is carried out as follows. Firstly, the function $S(\underline{Q})$ is measured directly at the reciprocal lattice points $\underline{\tau}$(hkl) in a diffraction experiment. Whereas its purely elastic part – in accordance with (4.53) – consists of a number of δ functions in $\underline{Q}$, its inelastic part is less structured in $\underline{Q}$ ("thermal diffuse scattering" = TDS). This can therefore be separated by graphic or numeric extrapolation in $\underline{Q}$, without it being necessary to analyse the energy of the scattered neutrons. In this sense it is justified to refer to diffraction sometimes as "elastic" scattering. If the TDS correction is significant (e.g. at higher temperatures), determining (4.53) by diffraction with energy analysis is however recommended. This is possible with neutrons (but not with X-rays). The well-known problem of diffraction on crystals will not be discussed further here. We refer to the pertinent literature /1.5,7;2.2/.

For the Van Hove self-correlation function at the limit $t\to\infty$ we find similarly (4.30)

$$G_s(\underline{r},\infty) = \frac{1}{N}\sum_{i=1}^{N}\int d\underline{r}'\,\langle\delta[\underline{r}+\underline{r}_i(0)-\underline{r}']\rangle\,\langle\delta[\underline{r}'-\underline{r}_i(t)]\rangle \quad . \tag{4.54}$$

If by $p_i(\underline{r}')d\underline{r}'$ we refer to the time-averaged probability that the ith atom is in the volume element $d\underline{r}'$ at point $\underline{r}'$ [$p_i(\underline{r}')$ is identical to $p(\underline{r}_0)$ of (4.32)]

$$p_i(\underline{r}')d\underline{r}' = \langle\delta[\underline{r}'-\underline{r}_i]\rangle\, d\underline{r}' \quad , \tag{4.55}$$

we have

$$G_s(\underline{r},\infty) = \frac{1}{N}\sum_{i=1}^{N}\int d\underline{r}'\, p_i(\underline{r}')\,p_i(\underline{r}-\underline{r}') \quad . \tag{4.56}$$

For systems in which the total (very large) volume V of the system is available for the motion of each atom (gases, liquids, etc.) clearly $p_i(\underline{r}')$ is negligibly small and constant. Then $G_s(\underline{r},\infty)$ contributes to the scattered intensity only at $(\underline{Q},\omega) = (0,0)$ and can therefore be neglected. More interesting is the case of a solid where the atom is bound to a restricted volume in the vicinity of one or several equilibrium positions, because there $p_i(\underline{r}')$ and thus $G_s(\underline{r},\infty)$ are generally $\neq 0$. Further, $G_s(\underline{r},\infty)$ can be regarded as a Patterson function of the individual atom and plays as important a part for incoherent scattering as $G(\underline{r},\infty)$ for coherent scattering. Similarly to $G(\underline{r},t)$ in (4.47), $G_s(\underline{r},t)$ can be separated into a time-independent and a time-dependent part

$$G_s(\underline{r},t) = G_s(\underline{r},\infty) + G_s'(\underline{r},t) \tag{4.57}$$

so that the first term corresponds to the elastic and the second term to the inelastic incoherent scattering. We therefore have

$$S_{inc}^{el}(\underline{Q},\omega) = \delta(\omega)\int e^{i\underline{Q}\underline{r}}\, G_s(\underline{r},\infty)d\underline{r} \quad , \tag{4.58}$$

$$S_{inc}^{el}(\underline{Q},\omega) = \delta(\omega)\, I_s(\underline{Q},\infty) \quad , \tag{4.59}$$

where because of (4.19,56)

$$I_s(\underline{Q},\infty) = \int e^{i\underline{Q}\underline{r}}\,\frac{1}{N}\sum_{i=1}^{N}\int d\underline{r}'\, p_i(\underline{r}')p_i(\underline{r}-\underline{r}')d\underline{r} \quad , \tag{4.60}$$

$$I_s(\underline{Q},\infty) = \frac{1}{N}\sum_{i=1}^{N}\left|\int d\underline{r}\, e^{i\underline{Q}\underline{r}}\, p_i(\underline{r})\right|^2 \quad . \tag{4.61}$$

In the case of a system of N equivalent atoms, the summation can again be omitted, so that

$$S_{inc}^{el}(\underline{Q},\omega) = \delta(\omega)\left|\int d\underline{r}\, e^{i\underline{Q}\underline{r}}\, p(\underline{r})\right|^2 \quad . \tag{4.62}$$

For the Q-dependent limiting value of the intermediate scattering function $I_s(\underline{Q},t\to\infty)$ the name "Elastic Incoherent Structure Factor" (EISF) has been adopted[6]. It contains the time-averaged probability density distribution $p(\underline{r})$ of the atoms and thus very important structural information. In proposing models for the incoherent scattering function one therefore generally starts from hypotheses for $p(\underline{r})$.

---

[6] A more detailed discussion of the EISF may be found in /4.9/.

5. Models for the Incoherent Scattering Function

To analyse scattering experiments models of atom movement are generally proposed and the corresponding scattering functions are derived. The values of the model parameters can be determined by fitting the theoretical formulae to the measured spectra. If the individual atom is subject to n different types of motion, and if it can be assumed that these are largely dynamically independent of each other, the incoherent scattering function $S_{inc}(\underline{Q},\omega)$ can be approximated by the convolution of the scattering functions for these different types of motion:

$$S_{inc}(\underline{Q},\omega) = \{\ldots\{S_1 \otimes S_2\} \otimes S_3 \ldots\} \otimes S_n \quad . \tag{5.1}$$

As an example, for three types of movement (e.g. translational jump diffusion S_{TS}, rotational jump diffusion S_{RS}, lattice vibrations S_{PH}) this formula is explicitly

$$S_{inc}(\underline{Q},\omega) = \int_{-\infty}^{\infty} d\omega' \int_{-\infty}^{\infty} d\omega'' \, S_{TS}(\underline{Q},\omega'') \cdot S_{RS}(\underline{Q},\omega'-\omega'') \cdot S_{PH}(\underline{Q},\omega-\omega') \quad . \tag{5.2}$$

In the following we assume that (5.1,2) are valid and discuss some of the simplest models. We shall sketch briefly the derivation of the scattering function for one example and then merely give the scattering functions for further examples.

5.1 Translational Diffusion

If the movement of the individual atom can be described by the macroscopic diffusion equation /5.1/, then

$$\frac{\partial}{\partial t} P(\underline{r},\underline{r}_0,t) = D^* \Delta P(\underline{r},\underline{r}_0,t) \quad , \tag{5.3}$$

with the initial condition $P(\underline{r},\underline{r}_0,0) = \delta(\underline{r})$; here D^* is the self-diffusion coefficient.

The solution of (5.3) is

$$P(\underline{r},\underline{r}_0,t) = (4\pi D^*|t|)^{-3/2} \exp\left[-(\underline{r}-\underline{r}_0)^2/4D^*|t|\right] \quad . \tag{5.4}$$

The probability density distribution $p(\underline{r}_0)$ of the starting points is in this case a trivial constant determined by the density of the sample. Its integral over the sample volume must obviously be equal to 1.

If (5.4) is inserted into (4.35), the classical incoherent scattering function for translational diffusion is found:

$$S^{cl}_{TD}(Q,\omega) = \frac{1}{\pi} \frac{D^*Q^2}{(D^*Q^2)^2+\omega^2} \quad . \qquad (5.5)$$

This is (for constant Q) a Lorentzian with the full energy width (FWHM) = $2\hbar D^*Q^2$ (full width at half maximum). By measuring this width as a function of Q the diffusion coefficient can be determined. This is always possible in principle, if D* is sufficiently large and if the measurement can be carried out at sufficiently low values of Q (rule of thumb: $Q \ll 2/a$, where a is the lattice constant of the crystal or the distance between nearest-neighbour atoms in the liquid). In the region of larger Q values the microscopic structure of the condensed system becomes important and (5.5) is no longer valid.

The simplest model for diffusion in solids that makes allowance for the structure is the translational jump model of CHUDLEY and ELLIOTT /5.2/. Here it is assumed that the diffusing atom moves in a regular lattice by random jumps between adjacent interstitial sites. The time required for the jump is neglected in comparison with the average residence time τ per site and it is assumed that the movements of different diffusing atoms are not correlated with each other. If the sites available for diffusion in the crystal form a Bravais lattice, the classical incoherent scattering function reads

$$S^{cl}_{TS}(\underline{Q},\omega) = \frac{1}{\pi} \frac{f(\underline{Q})/\tau}{[f(\underline{Q})/\tau]^2+\omega^2} \quad \text{with} \qquad (5.6)$$

$$f(\underline{Q}) = \frac{1}{n} \sum_{\nu=1}^{n} (1-\exp[-i\underline{Q}\underline{R}_\nu]) \quad , \qquad (5.7)$$

where $\underline{R}_\nu$ are the jump vectors leading to the n adjacent sites. As can be seen, this is again a Lorentzian, but this time with an oscillating energy width $2\hbar f(\underline{Q})/\tau$. The measurement of this width gives information not only on the rate $1/\tau$ of the diffusional jumps, but also on the geometry of the movement, via $f(\underline{Q})$. For more complex cases where the possible positions of the diffusing atoms do not form a Bravais lattice, a scattering function similar to (5.6) applies. This will not, however, be explicitly reproduced here. Mathematical details can be found in /5.3/.

5.2 Rotational Diffusion

A molecule in a condensed system on the time average having more than one well-defined orientation — apart from normal modes of small amplitude — is referred to as being in dynamic orientational disorder. The transition of the molecule to an adjacent orientation generally occurs by a statistical ("diffusion-type") rotational movement. In molecular liquids this type of movement is the rule, but it also occurs in certain molecular crystals /5.4/. The continuous rotational diffusion model is a simple model assuming equal probability for all orientations (i.e. spherical symmetry) on the time average.

This model corresponds to using the macroscopic diffusion equation (5.3) with the boundary condition that the motion of the diffusing atom remains limited to the surface of a sphere of radius a (distance of the atom from the center of gravity of the molecule). The corresponding classical incoherent scattering function is /5.5/

$$S_{RD}^{cl}(Q,\omega) = A_0(Q)\cdot\delta(\omega) + \sum_{\ell=1}^{\infty} A_\ell(Q)\cdot S_\ell(\omega) \quad , \tag{5.8}$$

where $A_\ell(Q) = (2\ell+1)j_\ell^2(Qa)$; $\ell = 0,1,2...\infty$; j_ℓ are spherical Bessel functions and $S_\ell(\omega)$ are Lorentzians of the form

$$S_\ell(\omega) = \frac{1}{\pi}\frac{\ell(\ell+1)D_r}{[\ell(\ell+1)D_r]^2+\omega^2} \tag{5.9}$$

with the full energy width $2\hbar\cdot\ell(\ell+1)D_r$; D_r is referred to as the rotational diffusion coefficient.

When the symmetry of the crystal lattice is of importance for the rotational motion, this results in an anisotropic time-averaged orientation distribution of the molecule. The simplest model which provides for such anisotropy is the rotational jump model /5.6-8/. Here it is assumed that the time-averaged orientation distribution is satisfactorily described by a finite number of discrete orientations. Accordingly, each atom of the molecule has a finite number of separate positions available on a spherical surface, between which it jumps when the molecule changes its orientation. If this model is considered from the viewpoint of the individual diffusing atom, the idea on which it is based – apart from the limiting number of positions – is identical with the above-mentioned Chudley-Elliot model. The associated classical incoherent scattering function is

$$S_{RS}^{cl}(\underline{Q},\omega) = A_0(\underline{Q})\delta(\omega) + \sum_j A_j(\underline{Q})L_j(\omega) \qquad \text{with} \tag{5.10}$$

$$A_0(\underline{Q}) = \frac{1}{n^2}\sum_{\nu=1}^{n}\sum_{\nu'=1}^{n}\exp[i\underline{Q}(\underline{R}_\nu-\underline{R}_{\nu'})] \qquad \text{and} \tag{5.11a}$$

$$L_j(\omega) = \frac{1}{\pi}\frac{\tau_j^{-1}}{(\tau_j^{-1})^2+\omega^2} \quad , \tag{5.11b}$$

where $\underline{R}_\nu$ are the position vectors of the n available sites. The coefficients $A_j(\underline{Q})$ are functions of these position vectors and $L_j(\omega)$ are Lorentzians whose widths are determined by the rotational jump rate $1/\tau$. Further details of this model and a number of other models of rotational molecular motion can be found in /5.9/.

5.3 Vibrations of Crystal Lattices and of Molecules

Atoms and molecules having fixed equilibrium positions and orientations in a crystal perform vibrations around these, the frequency spectrum of which is described by the phonon density of states $g(\omega)$. The wave-like propagation of these vibrations in the crystal lattice is investigated by phonon spectroscopy, which is the main application of coherent inelastic neutron scattering. As this topic is dealt with extensively in the literature /2.1;4.2,3;5.10,11/, we shall not go into further detail here.

In the simplest case (cubic crystal with only one atom per unit cell) incoherent neutron scattering permits direct experimental determination of $g(\omega)$ by measuring the incoherent scattering function for lattice vibrations:

$$S_{PH}(\underline{Q},\omega) = e^{-\langle u^2\rangle Q^2}\left\{\delta(\omega)+\langle u^2\rangle Q^2 S_1(\omega) + \frac{[\langle u^2\rangle Q^2]^2}{2!} S_2(\omega) + \ldots\right\} . \qquad (5.12)$$

This is an expansion of the scattering function by the number of phonon processes, where the first term describes purely elastic scattering, the second term single-phonon scattering, the third term two-phonon scattering, etc. The single-phonon expression is written explicitly

$$S_1(\omega) = g(\omega)\cdot\bar{n}(\omega) \quad , \quad \text{with} \qquad (5.13)$$

$$\bar{n}(\omega) = [2\omega\cdot\sinh(\hbar\omega/2k_BT)]^{-1}\cdot\exp(-\hbar\omega/2k_BT) \qquad (5.14)$$

and the condition that

$$\sum_{-\infty}^{\infty} S_1(\omega)d\omega = 1 \quad . \qquad (5.15)$$

Here $S_2(\omega)$ is the convolution of $S_1(\omega)$ with itself, and analogously the frequency-dependent factors $S_n(\omega)$ are the convolutions of $S_1(\omega)$ with $S_{n-1}(\omega)$. Here the "detailed balance" factor $e^{-\hbar\omega/2k_BT}$ appears explicitly in (5.14), since with the phonon scattering function S_{PH} the condition $|\hbar\omega| << k_BT/2$ is generally not fulfilled. If the primitive unit cell of the crystal contains more than one atom, the relationship between the density of states $g(\omega)$ and the scattering function $S_{PH}(Q,\omega)$ is no longer so simple; $g(\omega)$ must then be replaced by the effective density of states, which is the sum of the individual densities of states of the atoms in the unit cell weighted with the associated phonon polarisation vectors, reciprocal masses and Debye-Waller factors /1.1/.

It is not always necessary to use the complete phonon density of states to interpret measured spectra. Under certain conditions, particularly where because of weak coupling certain vibrational frequencies are especially strongly represented in the density of states, it appears justified to describe the dynamics of the system under investigation as an approximation by the dynamics of a harmonic oscillator with the eigenfrequency ω_0. The corresponding scattering function is /4.3/

$$S_{HO}(\underline{Q},\omega) = \exp\left[-\frac{\hbar Q^2}{2M\omega_0}\coth(\hbar\omega_0/2k_BT)\right]\exp(-\hbar\omega/2k_BT)\cdot\sum_{n=-\infty}^{\infty}\delta(\hbar\omega-n\hbar\omega_0)\cdot I_n(y) \quad , \tag{5.16}$$

where $I_n(y)$ is the modified Bessel function of the first kind, with

$$y = \frac{\hbar Q^2}{2M\omega_0}\operatorname{cosech}(\hbar\omega_0/2k_BT) \quad ,$$

M being the mass of the atom vibrating in an isotropic, harmonic oscillator potential.

5. 4 Convolution of Different Scattering Functions

At the beginning of Chap.5 we referred to the possibility – assuming dynamic independence of different types of motion – of obtaining the overall scattering function by folding the corresponding individual scattering functions. In this connection we now wish to point out an important qualitative difference between (5.5,6) on the one hand and (5.8,10,12) on the other. In contrast to the former, the scattering functions for rotational motions and lattice vibrations in principle always contain a purely elastic term, whose coefficient $A_0(\underline{Q})$ is the EISF defined in Sect. 4.3. Its appearance is due to the fact that the motion of the atom remains limited to a finite space of atomic dimensions. For example, if in addition to a rotational movement, translational diffusion according to (5.3) occurs, this spatial limitation is eliminated. The elastic term in the incoherent scattering function then disappears, as can be seen for instance by folding (5.5) with (5.8):

$$S^{cl}_{TRD}(Q,\omega) = A_0(Q)\cdot\frac{1}{\pi}\frac{D^*Q^2}{[D^*Q^2]^2+\omega^2}+\sum_{\ell=1}^{\infty}A_\ell(Q)S_\ell(\omega)\otimes S_{TD}(Q,\omega) \quad , \tag{5.17}$$

where $\otimes$ stands for the convolution of S_ℓ with S_{TD}. If there are two different types of motion which are both limited in space, the result of the convolution again contains an elastic term whose coefficient is the product of the EISFs of the two individual types of motion. An example is a system with lattice vibrations [according to (5.12)] and rotational jump diffusion (5.10). The folded incoherent scattering function reads

$$S_{RPH}(\underline{Q},\omega) = e^{-\langle u^2\rangle Q^2}\Big\{A_0(\underline{Q})\delta(\omega)+A_0(\underline{Q})\cdot\sum_{i=1}^{\infty}\frac{[\langle u^2\rangle Q^2]^i}{i!}S_i(\omega)$$

$$+\sum_j A_j(\underline{Q})\cdot L_j(\omega)+\sum_{i=1}^{\infty}\frac{[\langle u^2\rangle Q^2]^i}{i!}\sum_j A_j(\underline{Q})\cdot S_i(\omega)\otimes L_j(\omega)\Big\} \quad . \tag{5.18}$$

The EISF is given here by the product of the Debye-Waller factor $e^{-<u^2>Q^2}$ with $A_0(Q)$. The frequency dependence of the second term is determined by the one-phonon and multi-phonon functions $S_i(\omega)$, and that of the third term by the rotation functions $L_j(\omega)$. Finally in the fourth term the two spectral functions appear folded.

6. Specific Applications of Neutron Scattering

6.1 Localisation of Hydrogen Atoms in Crystals

In the presence of heavy atoms, light atoms, particularly hydrogen, can often be localised only inaccurately or not at all by means of X-ray diffraction methods, as
the scattering amplitude of an atom is proportional to the number of its electrons.
Heavy atoms therefore dominate in the scattered intensity.

In many cases neutron diffraction is considerably more advantageous, as the coherent scattering cross-sections of most atoms are of the same order of magnitude
(Table 3.1, Fig.3.1). Hydrate structures and hydrogen bonds have been investigated
for a long time by means of neutron diffraction /6.1/. Recently progress has been
made particularly in determining hydrogen positions in metal-organic complexes
/6.2/, biological structures /6.3/ and potential hydrogen storage materials. As an
example Fig.6.1 shows the structure of $H_{0.76}NbS_2$ /6.4/. In contrast to intercalation compounds (Sect.7.5) a hydride is formed here, as the hydrogen atom is in the
center between three niobium atoms. The niobium atoms are linked together in two-
dimensional layers and are "isolated" from each other by sulphur layers. As the hy-

◎ S at	2/3	1/3	0.37
	2/3	1/3	0.12
● Nb at	1/3	2/3	0.25
⊕ H at	0	0	0.25

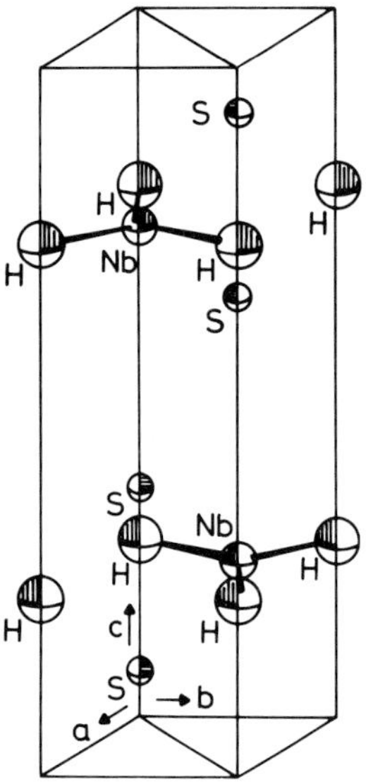

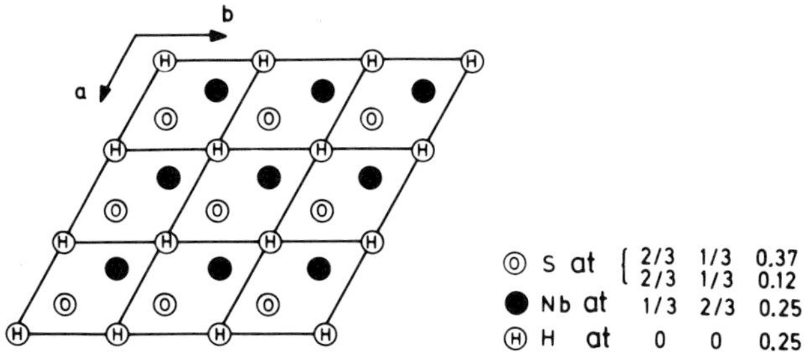

Fig.6.1. In the structure of $H_{0.76}NbS_2$ each hydrogen
atoms is surrounded by three niobium atoms. The atoms
are symbolised by anisotropic vibration ellipsoids.
To illustrate the two-dimensional character of the
structure the projection of a layer is shown in the
upper part of the illustration. The hydrogen positions
are only statistically occupied /6.4/

drogen content can be varied electrochemically over a wide range, this structure should be an interesting model system for investigating the two-dimensional transport of hydrogen. This system is also of interest in catalysis research, since the binding of hydrogen to catalysts based on MoS_2, which are of some importance in removing sulphur from crude oil, is not yet well understood /6.5/.

6.2 Investigation of the Chemical Bond with Neutron Diffraction

6.2.1 Combined Neutron/X-Ray Diffraction (X-N Method)

Whereas X-rays are scattered by the electron shells of the atoms, neutron scattering (apart from magnetic scattering) takes place at the atomic nuclei. This basic difference between the two methods is used to determine the distortion of the electron density distribution resulting from the chemical bonds in the crystal. First the precise positions of the atomic nuclei are determined by neutron diffraction. The neutron experiment also yields the Debye-Waller factor. This is very important, as it is often difficult (or impossible) to separate unambiguously Debye-Waller factors from X-ray form factors. With the aid of these data the structure factors of a pseudostructure, based on the X-ray form factors of spherical electron shells, can

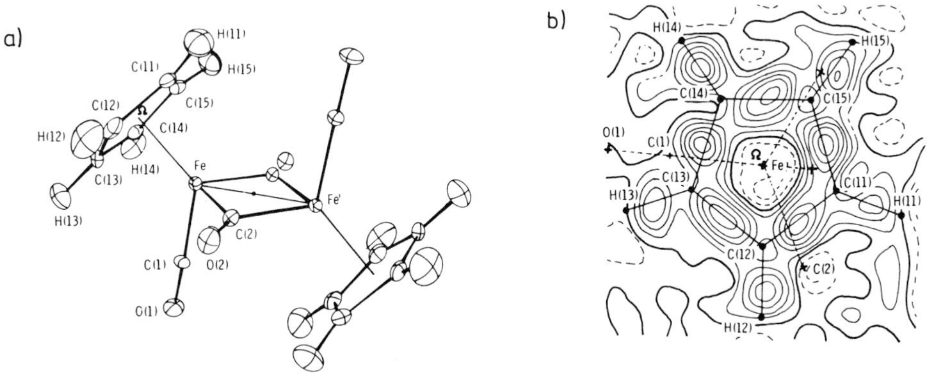

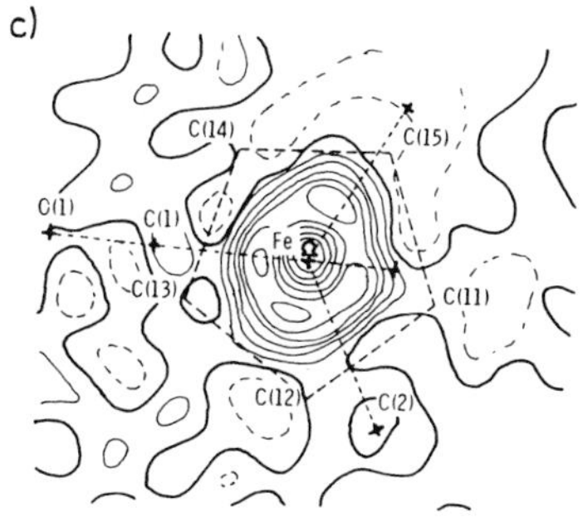

Fig.6.2. (a) Molecular structure of bis-(dicarbonyl-π-cyclopentadienyl)-iron. The atoms are symbolised by anisotropic vibration ellipsoids. (b) X-N deformation electron density: section at the level of the cyclopentadienyl ring. The non-spherical parts of the electron density on the lines connecting nuclei can be clearly seen. (c) Section at the level of an iron atom and parallel to the cyclopentadienyl rings /6.6/

be calculated (F_p^{hkl}). A further structure determination, this time with X-rays, gives the structure factors corresponding to the scattering from the actual electron distribution (F_r^{hkl}). The Fourier transform of ($F_r^{hkl}/K-F_p^{hkl}$) then yields the deformation electron density $\Delta\rho = \rho$ (crystal)-$\Sigma\rho$(pseudoatoms), which reproduces all the non-spherical contributions to the electron density. Here K is a normalisation factor.

The metal-organic compound bis-(dicarbonyl-π-cyclopentadienyl)-iron (trans-$[(\pi-C_5H_5)Fe(CO)_2]_2$, Fig.6.2a /6.6/) may be mentioned here as an example. The deformation electron density at the level of the cyclopentadienyl ring clearly shows the bonding electrons on the C-C and C-H connecting lines as expected according to the usual models of the chemical bond (Fig.6.2b). A further section (Fig.6.2c), this time at the level of the iron atom and parallel to the aromatic ring, shows that the Fe atom is surrounded by electron density in the form of a ring, which indicates an uneven population of the d orbitals. This experimental result is supported by a semi-empirical molecular orbital calculation (SCCCMO) /6.6/. For a review of the X-N method, see /1.7;6.7/.

6.2.2 Determination of the Spin Density with Polarised Neutrons

Neutrons possess a magnetic moment and are scattered by interaction with the magnetic moments of unpaired electrons. In the case of ions of 3d-transition metals, these are essentially the magnetic spin moments of the electrons. (The magnetic orbital moments are largely suppressed by interaction with the adjacent atoms.) Paramagnetic ions with disordered electron spins lead to diffuse scattering. Discrete magnetic reflections are, however, observed for ordered spin structures /1.7;2.2/. With ferromagnetic compounds only parallel spins occur, so that nuclear scattering and magnetic scattering are superimposed. A separation is possible by subtracting the spectrum measured above from that measured below the ordering temperature (Curie temperature). With antiferromagnetic substances there may be a difference between the crystallographic and the magnetic unit cell. There is a particularly simple case where the spins of neighbouring, translationally equivalent atoms are arranged in antiparallel fashion. If this is the case, for example along the three axes of a cubic unit cell, the volume of the resulting magnetic elementary cell increases by a factor of eight in comparison with the crystallographic unit cell, and superlattice reflections are observed.

Knowing the magnetic structure can contribute to the understanding of chemical bonding. A well-known example is the "super exchange", which in the case of paramagnetic cations, connected via diamagnetic anions (e.g. FeO), leads to an antiferromagnetic ordering /1.7/. This can be explained by a partial electron exchange between cations and anions. For example, if a fully occupied anion p orbital overlaps with one d orbital of each of the two adjacent cations, antiparallel spins are induced on the cations as long as these show the same electron structure, in order not to violate the Pauli principle for the p orbital of the anion.

More insight into chemical bonding, as far as it concerns unpaired electrons, can be obtained from a spin density distribution, which – in analogy with the electron density distribution – can be calculated by a Fourier transformation of the magnetic structure factors (F_M^{hkl}). However, the intensities of most of the magnetic reflections in comparison with the crystallographic Bragg reflections are too weak for them to be determined by unpolarised neutrons. This is mainly due to the large spatial extension of the probability density distribution of the unpaired electrons, which causes – even more than for X-ray scattering – a decrease of the scattered intensity with increasing scattering angle. Polarised neutrons are considerably better suited for measuring magnetic scattering. This will be discussed for a ferromagnetic or paramagnetic compound, whose spins are oriented perpendicular to the scattering plane by means of a magnetic field. This is shown schematically in Fig. 6.3. Here $\underline{H}$ is the unit vector in the direction of the magnetic field, $\underline{\mu}(s)$ is the magnetic spin moment, and s is the effective spin quantum number of the atom or ion concerned. The two diagrams of Fig.6.3 differ by the neutron polarisation directions $\uparrow$ and $\downarrow$. It can be shown that the intensities of a reflection are different for the two polarisation directions:

$$\uparrow I^{hkl} \propto (F^{hkl} + F_M^{hkl})^2 \quad , \tag{6.1a}$$

$$\downarrow I^{hkl} \propto (F^{hkl} - F_M^{hkl})^2 \quad . \tag{6.1b}$$

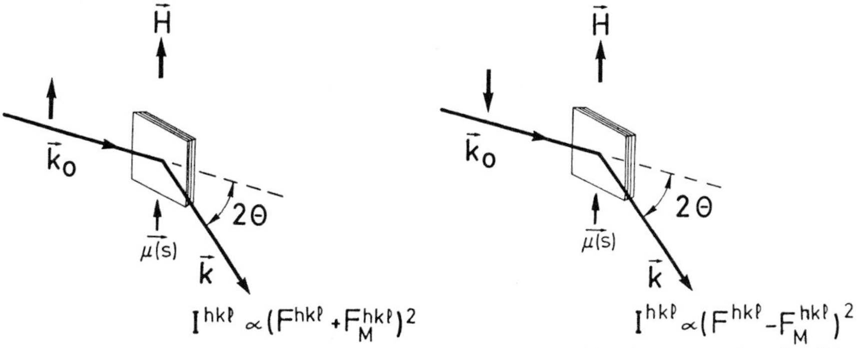

Fig.6.3. Schematic representation of a neutron scattering experiment with polarised neutrons on a ferromagnetic crystal. Neutrons with a defined polarisation direction ($\uparrow$ or $\downarrow$) are scattered. The direction of the magnetic field is indicated by the unit vector $\underline{H}$, and the magnetic spin moments by $\underline{\mu}(s)$. $\underline{H}$ and $\underline{\mu}(s)$ are perpendicular to the scattering plane defined by the neutron wave vectors $\underline{k}_0$ and $\underline{k}$. For the polarisation direction $\uparrow$ the scattered intensity is $I \propto (F+F_M)^2$, and for $\downarrow$ it is $I \propto (F-F_M)^2$, where F and F_M are the structure factors for nuclear and magnetic scattering, respectively

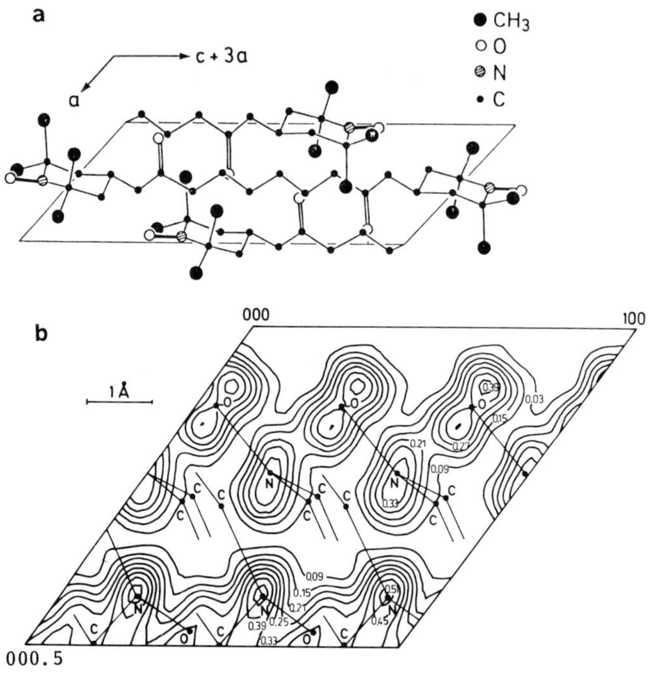

Fig.6.4. (a) Projection of the structure of the organic diradical d-(2,2,6,6-tetramethyl-4-piperidyl-1-oxyl)-suberate on the ac plane /6.8/. (b) Projection of the spin density of the molecule shown in (a) in direction $[\bar{4}3\bar{3}]$, i.e. perpendicular to the p_π-p_π bond of the NO group. Adjacent lines differ by 0.06 $\mu_B/\text{Å}^2$ (μ_B = Bohr magneton)

The ratio $\uparrow I^{hkl}/\downarrow I^{hkl}$ can be determined much more precisely than the ratio $(F^{hkl})^2/[(F^{hkl})^2+(F_M^{hkl})^2]$, which is obtained in measurements with unpolarised neutrons below and above the Curie temperature. To separate F and F_M an additional independent determination of the structure factors with X-ray or neutron diffraction is necessary above the Curie temperature.

The spin density determination of a stable organic diradical, d-(2,2,6,6-tetramethyl-4-piperidyl-1-oxyl)-suberate with the molecular formula $C_{26}H_{46}N_2O_6$ (Fig. 6.4a), is mentioned here as an example /6.8/. This compound has a Curie temperature of 0.38 K. In the molecular orbital description the unpaired electron of the NO group is in an antibonding orbital formed by the 2p orbitals of nitrogen and oxygen.

Ab initio calculations on the nitroxyl radical (H_2NO) led to contradictory results, as they localised the unpaired electron on either the nitrogen or oxygen atom /6.9,10/. The measurement with polarised neutrons was carried out at 1.5 K, the sample being in a 1.65 kOe magnetic field. The projection of the spin density on to a plane parallel to the p_π orbital is shown in Fig.6.4b. It can be seen that the spin density reaches a maximum value perpendicular to the N-O connecting line and a minimum value between N and O, as is to be expected for an antibonding wave function of 2p orbitals. However, in contrast to the theoretical predictions, it is found that the spin density is localised approximately equally on the nitrogen and oxygen atoms.

No classical structure determination can be carried out on many synthetic and natural macromolecules, as they cannot by crystallised. Indications of size and spatial distribution of the molecules can, however, also be obtained from the analysis of diffuse scattering at small diffraction angles. In particular this method permits the study of macromolecules in solution.

In this section we limit the discussion to the region of very small Q values, the so-called Guinier region, $0 < Q \cdot R_G < 1$ /6.11/. The practically attainable lower limit of Q is determined by the experimental technique and is in the most favourable case near $Q \cong 10^{-3}$ Å^{-1}. In this region the radius of gyration $\langle R_G^2 \rangle^{1/2}$ of the macromolecule can be determined[7].

The quantity R_G^2 (6.2) is the mean square of all the atomic distances from the center of gravity of the scattering molecule, weighted with the scattering lengths b_i. According to the Lagrange theorem R_G can be expressed by the m interatomic distances r_{ij} as follows:

$$R_G^2 = \frac{1}{2} \cdot \sum_{i,j} b_i b_j r_{ij}^2 / \sum_{i,j} b_i b_j \quad . \tag{6.2}$$

The shape of the particles generally influences the character of the small-angle scattering curve. In the region of very small scattering vectors, however, the particle shape can be neglected, and we obtain R_G^2 directly from the scattering curve by extrapolation of the intensity towards $Q = 0$ /6.11/.

$$I(Q) = I_{Q=0} \exp(-R_G^2 Q^2/3) \quad . \tag{6.3}$$

Here a strongly dilute solution is assumed. For concentrated solutions and even solids the scattering of the single molecule can also be determined, if some of the macromolecules are deuterated. This method corresponds to the heavy-atom substitution technique used in X-ray small-angle scattering. It is, however, an essential advantage of the deuteration method used in neutron scattering that the H/D substitution generally alters intermolecular interactions so little that the change can be neglected.

6.3.1 Polymers

In the discussion of the spatial distribution of polymer molecules we first consider the basic model of a linear polymer with n mobile chain segments. Here a Gaussian

[7] For brevity and when no confusion would arise, we shall write R_G^2 instead of $\langle R_G^2 \rangle$ and R_G instead of $\langle R_G^2 \rangle^{1/2}$ in the following.

statistical distribution applies, for which the expression "Gaussian coil" is used. This model is realised in the amorphous solid /6.12/. For dissolved polymers, the Gaussian coil is present only in the so-called Θ solution, in which the attraction and repulsion of the chain segments, mediated by the interaction with the solvent, cancel each other. Poor solvents, on the other hand, cause the collapse and good solvents the swelling of the dissolved molecules (Sect.7.6). These three conformations (i.e. Gaussian, swollen and collapsed coil) differ with respect to the scaling of the radius of gyration as a function of the molecular weight (M_w) /6.13/:

Gaussian coil: $R_G \propto M_w^{1/2}$ (6.4)

Swollen coil: $R_G \propto M_w^{3/5}$ (6.5)

Collapsed coil: $R_G \propto M_w^{1/3}$ (6.6)

As R_G can be determined by small-angle scattering experiments [cf. (6.3)], this gives a simple possibility of obtaining information on the conformation of molecules. This may be demonstrated by the example of deuterated polystyrene with the monomer unit

$$\left[-CD_2 - \overset{\overset{\textstyle C_6D_5}{\textstyle |}}{CD} - \right]_n \quad .$$

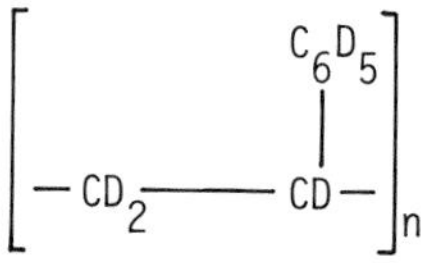

Solutions in a good solvent (carbon disulphide-CS_2) and in a Θ solvent (cyclohexane-C_6H_{12}, at 36°C), as well as the amorphous solid /6.14/ were investigated. Figure 6.5 shows that the scaling (6.4) corresponding to the Gaussian coil is indeed observed for the polymer dissolved in C_6H_{12} and for the amorphous solid. As was to be expected, systematically larger R_G values are found in the good solvent, where the swelling leads to an expansion of the spatial distribution of the molecules.

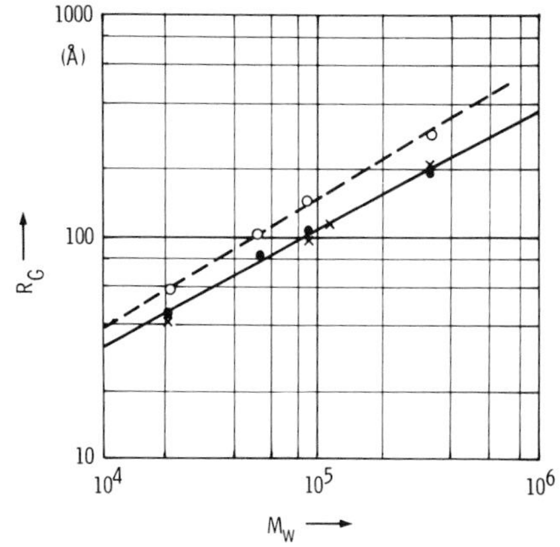

Fig.6.5. Small-angle scattering on dissolved and amorphous polystyrene /6.14/. For a Θ solution (cyclohexane at 36°C; ×) the radius of gyration R_G and the molecular weight M_W scale in the same way as for the amorphous polymer (●). According to model predictions, this corresponds to the Gaussian coil (see text). The systematically larger R_G values in a good solvent can be explained by a swelling of the chains (o)

By changing the concentration or temperature it is possible to influence the conformation of the dissolved polymers. Under certain conditions transition phenomena appear in which the molecules can no longer be described by the simple models mentioned above. Here scattering experiments beyond the Guinier region may aid analysis (Sect.7.6.1).

6.3.2 Biological Macromolecules

The Guinier region is also important when investigating dissolved biological macromolecules /6.15,16/. Such molecules often form complex associations "in vitro", which can be studied by neutron small-angle scattering experiments on "in vivo" model systems. If different molecules are present in the same solution, the following equation applies for the effective radius of gyration

$$R_G^2 = \left[\sum_i c_i A_{i,Q=0}^2 R_{G,i}^2\right] / I_{Q=0} \quad , \tag{6.7}$$

where c_i is the molar concentration, $R_{G,i}$ the radius of gyration and A_i the structure factor of the ith kind of molecule (at $Q = 0$ this is simply the sum of scattering lengths) and $I_{Q=0} = \sum_i c_i A_{i,Q=0}^2$. When calculating $A_{i,Q=0}$ it is necessary to take into account the part of the volume (V_i) of the solvent replaced by the molecule:

$$A_{i,Q=0} = \left(\sum_j b_j\right)_i - \zeta_s V_i \quad , \tag{6.8}$$

where ζ_s is the mean scattering length density of the solvent. Some of the hydrogen atoms of biological macromolecules are exchangeable. Thus the scattering density of the molecule can be adapted to that of the solvent by the appropriate choice of the H_2O/D_2O ratio in the solvent $[(\sum_j b_j)_i = \zeta_s V_i]$.

This is called masking. Different types of biological macromolecules, such as proteins or nucleic acids, generally contain different numbers of replaceable protons. This can be utilised when associations of two types of molecules are present. If the appropriate H_2O/D_2O ratio is chosen, one type of molecule is selectively masked and the small-angle scattering can be interpreted for the second type of molecule (contrast variation technique).

This may be exemplified by the associations of a transfer ribonucleic acid (tRNA) with an aminoacyl-tRNA-synthetase (RS) /6.17/.

The synthetase catalyses "in vivo" the transfer of an amino acid to tRNA molecules. A specific tRNA molecule exists for each amino acid. This is one step of the protein synthesis which takes place at the ribosomes. In a preceding step the RNA-polymerase catalyses the formation of a single-strand copy of the two-stranded nucleic acid (double helix). This copy known as "messenger" ribonucleic acid (mRNA)

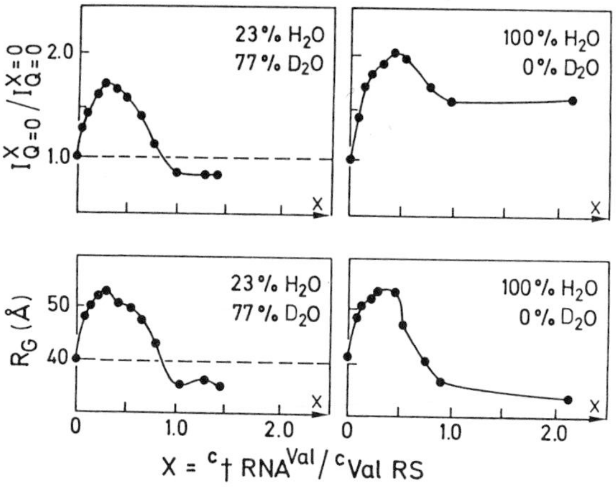

Fig.6.6. Small-angle scattering on biological macromolecules. The intensity extrapolated to $Q=0$ ($I^X_{Q=0}$), and the mean gyration radius (R_G) were determined for various concentration ratios $X = c_{tRNA^{Val}}/c_{ValRS}$ of mixtures containing an enzyme (ValRS) and a nucleic acid ($tRNA^{Val}$). $I^X_{Q=0}$ was normalised to the scattering of a pure ValRS solution. The continuous lines refer to model calculations mentioned in the text and support the idea of concentration-dependent association equilibria /6.17/

binds itself to the ribosome. Here the base pairing of the tRNA molecules with the mRNA molecule takes place, so determining the protein sequence.

The association between valyl RNA synthetase (ValRS, $M_W = 170000$) and the tRNA molecule ($tRNA^{Val}$, $M_W = 25000$) transmitting the amino-acid valine was investigated by neutron scattering /6.17/. To study the effect of the concentration of molecules on the degree of association, the ratio $X = c_{tRNA^{Val}}/c_{ValRS}$ was varied. Figure 6.6 shows $I^X_{Q=0}$ (normalised to $I^{X=0}_{Q=0}$) and R_G for solutions in pure water and in 23% H_2O/ 77% D_2O. The $tRNA^{Val}$ is selectively masked for the isotope mixture. Let us first consider this solution. For small X values $I^X_{Q=0}$ and R_G increase until a maximum is reached at $X \sim 0.5$. With increasing X values both $I^X_{Q=0}$ and R_G diminish. This indicates an association of several synthetase molecules for $X \leq 0.5$, and can be described by the following equilibria:

$$E + N \overset{k_1}{\rightleftarrows} EN \quad , \tag{6.9a}$$

$$E + EN \overset{k_2}{\rightleftarrows} E_2N \quad , \tag{6.9b}$$

$$E + E_2N \overset{k_3}{\rightleftarrows} E_3N \quad , \tag{6.9c}$$

$$E + E_3N \overset{k_4}{\rightleftarrows} E_4N \quad , \tag{6.9d}$$

(E = enzyme, N = nucleic acid).

This is the simplest system of equations for which a quantitative description of the $I^X_{Q=0}$ variation can be obtained (full curves in Fig.6.6; cf. /6.17/ for the determination of the equilibrium constants). According to biochemical investigations, multi-enzyme complexes probably also exist "in vivo" /6.17/. For X = 1 the equilibrium is displaced to the (1:1) complex. Here R_G diminishes from ≅ 39 Å to ≅ 35 Å, indicating a considerable change in the spatial distribution of the synthetase in its association with the nucleic acid. As furthermore a reduction of $I_{Q=0}$ by ≅ 15% is found in the (1:1) complex, it must be assumed that the specific volume of the synthetase is reduced by ≅ 1%. In pure water on the other hand, the first term of (6.8) is considerably larger than the second, so that an increase of $I_{Q=0}$ is found for the (1:1) complex. Finally the diminution of R_G in pure water should be mentioned. As the tRNA molecule here contributes to the small-angle scattering, it must be assumed that it is located in the vincinity of the center of mass of the complex.

The present example shows that complex association equilibria can be studied by means of a relatively simple experiment and with little expenditure in the analysis. However, the possibilities of small-angle scattering go further, as information on the shape of the molecules can be obtained beyond the Guinier region. A survey of these applications is given in /6.15,16/.

6.4 Identification of Mobile Molecular Groups Using Isotope Substitution

The isotope substitution method can be used to study the relative mobilities of different parts of a molecule. For this purpose it is sufficient to compare the neutron spectra of the various derivatives of the same substance obtained by isotope substitution. Direct conclusions can be drawn from the spectral differences on the contributions of different atoms or molecular groups to the scattered intensity. This method is particularly advantageous for materials containing hydrogen, where the great difference between the incoherent scattering cross-sections of the isotopes H and D can be utilised (cf. Table 3.1 and Fig.3.1).

A suitable example is the investigation of the methyl group rotation in the organic crystal para-azoxy anisole (PAA) /6.18/. In the temperature range 119°C ≦ T ≦ 135°C PAA has a liquid-crystalline (nematic) phase, in which the molecules perform diffusive translational and rotational motions. Additional internal rotational movements of molecular end groups can be observed only with difficulty in the nematic phase, as they are superimposed on the other types of motion. However, in the solid phase (T < 119°C) the molecular centers of gravity are fixed, and it is of interest to know to what extent this also applies for the orientation of molecules and molecule end groups. Measuring the inelastic spectra of the two partially deuterated PAA derivatives PAA-ϕD4 and PAA-CD3 answers this question (Fig.6.7). Due to the dominating incoherent scattering cross-section of the hydrogen atom the spectra of PAA-ϕD4 mainly contain information on the movement of the two methyl groups, whereas the spectra of PAA-CD3 are characterised by the movement of the phenyl groups. Comparing these spectra therefore permits the determination of the spectral contributions from the various parts of the molecule and thus the separation of those parts

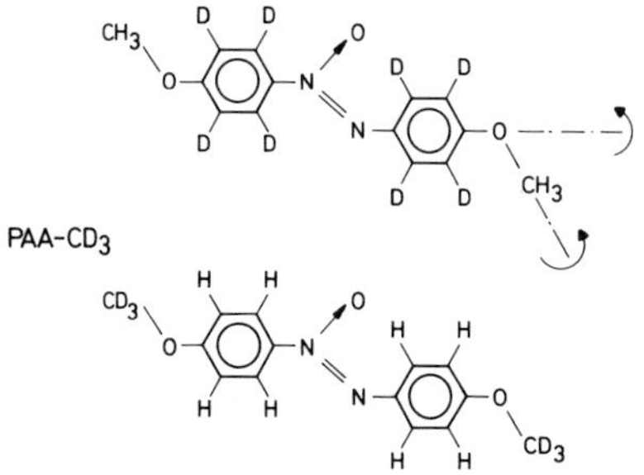

Fig.6.7. Partially deuterated forms of para-azoxyanisole (PAA). Deuteration of the phenyl rings: PAA-ϕD4; deuteration of the methyl groups: PAA-CD3

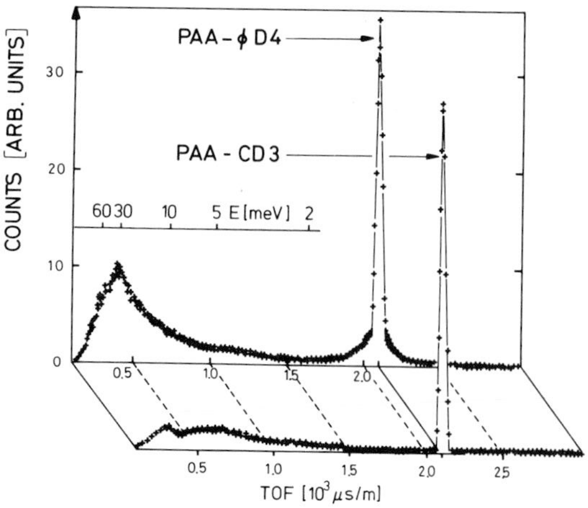

Fig.6.8. Comparison of the neutron time-of-flight spectra of PAA-ϕD4 and PAA-CD3 /6.18/. Temperature: 100°C; wavelength of the incident neutrons: λ_0 = 8.25 Å; scattering angle: φ = 125°; elastic energy resolution: FWHM = 0.048 meV

of the scattered intensity due to the methyl groups and to the rest of the molecule, respectively. This is shown in Fig.6.8 with the aid of time-of-flight spectra of the two PAA derivatives. The PAA-ϕD4 spectrum is characterised by two dominant features, which completely disappear when the methyl groups are deuterated: a) quasielastic scattering under the (well-resolved) elastic peak and b) strong inelastic scattering with a maximum close to $\hbar\omega$ = 31 meV. It can immediately be concluded from these two observations that the methyl groups perform both diffusive motions of considerable amplitude (quasielastic scattering) and periodic high-frequency movements (31 meV), and that the phenyl groups do not participate in either of these movements. The presence of the elastic components shows that the diffusive motion remains limited to a finite volume of molecular order of magnitude, and suggests that this is a rotational motion. This suggestion was confirmed by comparison of the spectra with model calculations. We shall go into this in more detail in the next section.

As a further example of the successful use of isotope substitution we should also like to mention a study of polymer dynamics. According to the rotational isomer model /6.12/, polymer molecules in the amorphous state and in solution take on innumerable different conformations, which arise due to internal rotations about bonds in the main chain or by rotation of side groups. In the liquid state (melt or solution) the movement of the chain molecule can be described by translational diffusion of its center of gravity, while the molecule itself is continuously changing its configuration. In rubber-type polymers the center-of-gravity diffusion is rather slow, but internal rotational motions can be very fast, so that it is possible for a molecule to go through the numerous different rotational isomers in a short time. The flexibility of the main chain, based on internal rotations, disappears only at the glass transition. Merely the rotation of small side groups continues to exist in some cases, until it "freezes" at lower temperatures. A detailed microscopic description of the complete dynamics of amorphous, liquid or dissolved polymers is of course possible only with difficulty because of the lack of symmetry and the resulting complexity of these systems (Sect.7.6). Specific experiments can, however, solve important parts of the problem, such as the question of the relative mobilities of main chains and side groups of polymer molecules as a function of temperature. Thus the correctness of the model just described, where the glass transition is considered to be a transformation from a phase with flexible ("wriggling") chains to a phase with more or less immobile ("frozen") chains, was proved by incoherent neutron scattering associated with isotope substitution.

A polymer, polypropylene oxide, with the monomer unit

(a)

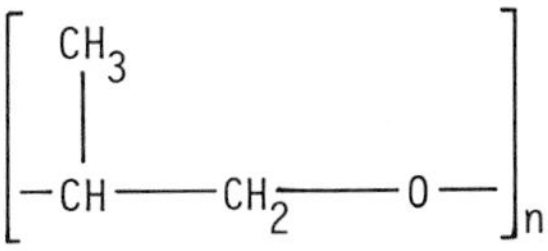

was investigated in this form and in two selectively deuterated forms, (b) methyl group deuterated, (c) main chain deuterated /6.19,20/. In the glass phase ($T < 200$ K) only pure elastic scattering was found for molecules of type (b), whereas (a) and (c) gave incoherent spectra, each with an elastic and a quasielastic component. This and the weak Q dependence of the measured quasielastic linewidth permit the unequivocal interpretation of the quasielastic component as due to a rotational movement of the methyl groups. The observed rotation "frequencies" are in the range of 10^{10} Hz and their temperature dependence corresponds to an activation energy of 17 kJ/mol.

Above the glass transformation quasielastic scattering was observed on the type (b) polymer, but not on type (c). It is therefore certainly attributable essentially to the movement of the main chain. The quasielastic linewidth shows the strong Q dependence typical for the segment motion of polymer chains (Sect.7.6.2), further indicating the correctness of this explanation.

6.5 Determination of Molecular Rotation Radii, Axes of Rotation and Rotation Rates

We now return to the example of PAA discussed in the previous section. The structure of the PAA molecule permits the following plausible models to be envisaged (Fig.6.7):

A. The O-CH_3 group rotates about the ϕ-O bond;

B. The CH_3 group rotates about its C_3 axis;

C. Both types of rotation are permitted.

The dependence of the purely elastically scattered intensity on the scattering vector Q (Sect.4.4: EISF) is via Fourier transformation a morror image of the trajectory of the scattering particle – in this case the hydrogen atom. The rotation radius can be determined directly from this measured function by comparison with the different calculated model functions (Fig.6.9). In this way it was possible to make a clear decision giving priority to Model B /6.18/. This combination of partial deuteration with the measurement of the EISF has also proved very effective to identify rotating molecular groups in other cases. Thus in 1974 it was possible for the first time to prove the rotation of the TBBA molecule (Fig.6.10) (terephthal-bis-butyl-aniline) about its long axis in the smectic H-(B) phase of this liquid crystal /6.21/. With this and subsequent publications /6.22,23/ the authors succeeded in terminating a controversy on the nature of molecular ordering in smectic phases.

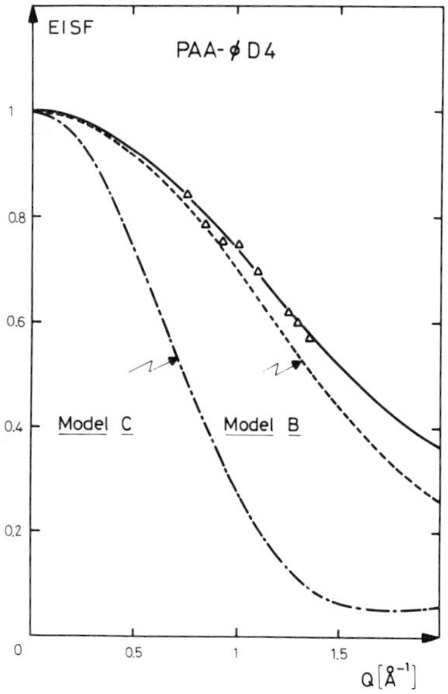

◄ Fig.6.9. Elastic incoherent structure factors (EISF) for PAA-ϕD4: (Δ) measurement; (·—·—·) Model C (simultaneous rotation of the CH_3 and $O\text{-}CH_3$ groups); (-----) Model B (rotation of the methyl groups alone); (——) Model B, corrected for imperfect deuteration (95% D and 5% H in the phenyl rings /6.18/

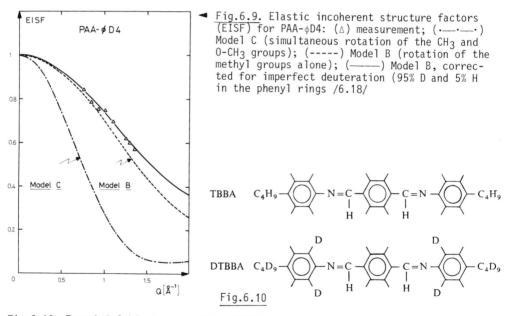

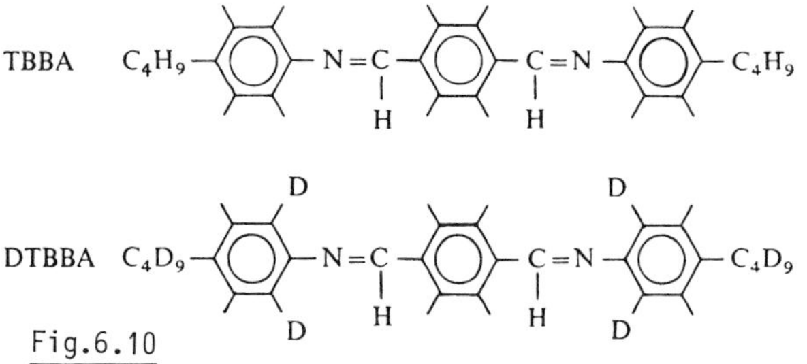

Fig.6.10

Fig.6.10. Terephthal-bis-butyl-aniline (TBBA) and a partially deuterated derivative (DTBBA) of it

In the field of molecular crystals with orientational disorder (ODIC), in which the molecules have no fixed orientation ("plastic" crystals and rotator phases /5.9/), but rotate back and forth between different minima of the orientation potential ("rotational jumps"), one of the most important problems is the determination of the heights of the potential barriers between these minima. In Fig.6.11

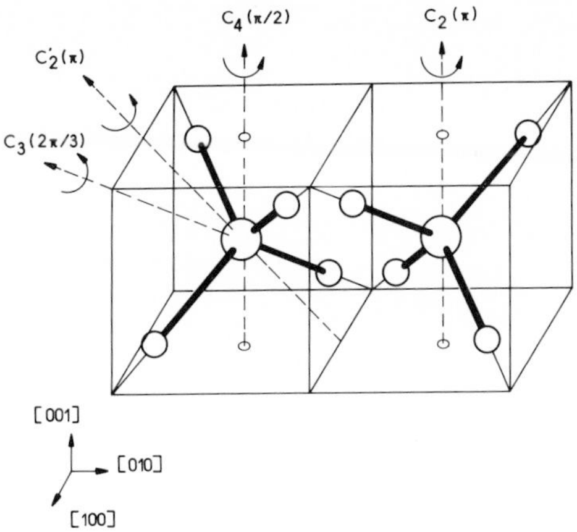

Fig.6.11. Schematic representation of the two possible orientations of the NH_4^+ group in the cubic CsCl phase of NH_4Cl, together with the four different axes of rotation discussed in the text

this problem is explained with the aid of the rotator phase of an ammonium halogenide, e.g. NH_4Cl (Phase II). The structure is of the cubic CsCl type, in which the ammonium ions are randomly distributed over the two different orientations relative to the cubic lattice shown (schematically) in Fig.6.11 [8]. The transition of an ammonium group from one orientation to another may take place by means of one of the class C_2' rotations shown in Fig.6.11 (180° rotation about one of the six twofold axes [110], [101], [011], etc.) or by a C_4 rotation (90° rotation about one of the three fourfold axes [100], [010], [001]). Rotations of type C_2 and C_3, on the other hand, produce transitions between indistinguishable orientations of the tetrahedron. In addition all the possible combinations of the four rotation classes mentioned are of course conceivable (a total of 16). With the aid of incoherent neutron scattering it is possible to distinguish between these different types of movement and thus to give a qualitative indication of the heights of the potential barriers corresponding to the various rotation axes. Thus measurement of the EISF of plastic adamantane ($C_{10}H_{16}$, face-centred cubic structure) showed that at room temperature C_4 and/or C_2' rotations of the molecules are dominant and that C_2 and/or C_3 rotations certainly do not occur alone, if they occur at all /6.24/. This result is illustrated in Fig.6.12 by comparing calculated EISF model curves with the values measured on adamantane powder. These measurements reduce the number of possible combinations of different rotation classes to 12, each of which contains either class C_4 or C_2' or both. It is not possible to distinguish by the elastic incoherent scattering alone between these twelve combinations. Measurement of the $\underline{Q}$ dependence of the elastic *and* quasielastic scattering on a single crystal, however, enables further progress. In this way, for example, it was possible

[8] The question whether the orientations of adjacent ions or molecules are correlated over a certain distance (in the order of magnitude of a lattice constant) must presumably be answered positively. However, at temperatures not too close to the phase transition, such correlations are probably so weak that the approximation of a random distribution of orientations is sufficient in our context.

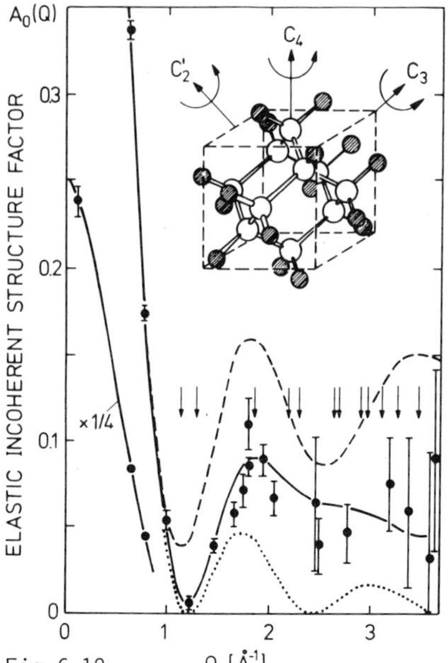

Fig.6.12

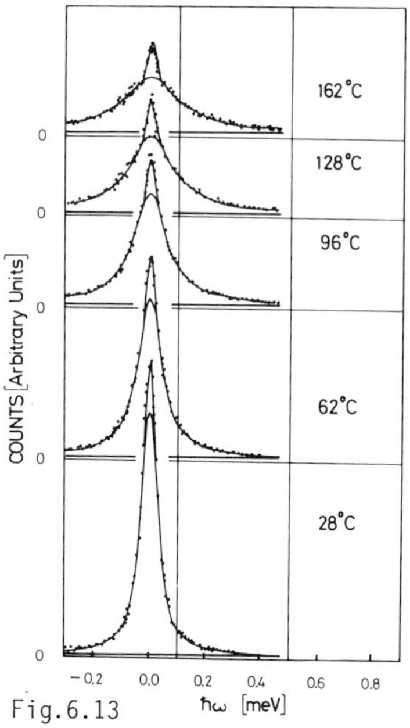

Fig.6.13

Fig.6.12. EISF of polycrystalline adamantane, $C_{10}H_{16}$: (⧳) measured values /6.24/ with statistical error bars; (.....) isotropic rotational diffusion; (-----) 120° rotational jumps about threefold axes of symmetry of the lattice; (———) 90° rotational jumps about fourfold axes of symmetry; the vertical arrows indicate the positions of the Bragg reflections of adamantane

Fig.6.13. Separation of the neutron spectra /6.28/ of adamantane (at 5 different temperatures) into the purely elastic, quasielastic and inelastic components. (Wavelength of the incident neutrons: λ_0 = 9.494 Å; scattering angle: φ = 100.8°; elastic energy resolution: FWHM = 0.031 meV.) The elastic component (narrow maximum) has a width independent of temperature, which is determined by the experimental energy resolution of the scattering facility. However the quasielastic component observed underneath the elastic peak shows considerable broadening with increasing temperature. The weak, very flat ("inelastic") contribution of the phonons in the quasielastic region does not change its basic form with temperature, but increases slowly in intensity

to show the presence of both C_4 and C_3 rotations in NH_4Cl-II /6.25/, the rotator potential being qualitatively similar to that of adamantane. At the II/III phase transition (243 K) both take place with a rate of about 10^9 s^{-1} (corresponding to an average residence time per orientation of $\tau_{C_3} \cong \tau_{C_4} \cong 10^{-9}$ s; Chap.5). On transition to Phase III ($T < 243$ K), τ_{C_4} increases by a factor 4, whereas τ_{C_3} essentially does not change[9]. In adamantane on the other hand, we were recently able to show that

[9] It was recently found in computer experiments /6.26/ that a considerable proportion of these C_3 rotations occur by multiple C_4 rotational jumps without stop in between jumps. In the context of the rotational jump model the latter are equivalent to the C_3 rotations.

for residence times $\tau < 5\cdot10^{-11}$ s only rotations of class C_4 occur, so that on this time scale all other rotation combinations can be excluded /6.27/.

This is a positive proof that in adamantane the potential barriers corresponding to the fourfold rotation axes [100], [010], and [001] are the lowest ones. The quantitative determination of this barrier in the sense of an activation energy can be carried out by measuring the temperature dependence of the quasielastic linewidth (Fig.6.13). The following Arrhenius law was obtained for the reciprocal residence time (rotational jump rate) /6.28/:

$$1/\tau_{C_4} = 5.2\cdot10^{12}\cdot\exp(-1350/T[K])\ s^{-1} \quad .$$

6.6 Rotational Tunneling in Molecular Crystals

The phenomenon of random reorientation of molecules or molecular groups (rotational jumps) discussed in the last two sections corresponds to the classical behavior which occurs at sufficiently high temperatures. By coupling to the crystal lattice vibrations (phonons) the molecules are incited from time to time to jump over the rotation potential barriers lying between fairly well-defined "discrete" orientations. This phenomenon is apparently based on statistical fluctuations both in the momentary height of the potential barriers and the momentary rotation energy of a molecule about the corresponding time averages. At low temperatures, when these fluctuations become small, sharp lines are observed at $\hbar\omega \neq 0$, corresponding to the transitions between well-defined, discrete energy levels, instead of the quasielastic scattering centred at $\hbar\omega = 0$. The classical description of the motion must then be replaced by the quantum-mechanical treatment.

With low potential barriers V_M (i.e. when $V_M \ll B = \hbar^2/2I$; I = moment of inertia of the molecule) the rotation is nearly free. This is known to be the case only for solid hydrogen. Almost free rotation is, however, also observed when V_M is of medium size and at the same time the rotation potential is highly symmetric. This was found for example in solid methane below the I/II phase transition ($T_{I/II} = 20.4$ K) /4.6; 6.29/. There are indications that Phase II of CH_4 — similar to CD_4-II — has a cubic structure with "antiferro" order and eight molecules per primitive elementary cell. While six of these molecules have a fixed orientation, the remaining molecules rotate almost freely. The energy eigenvalues for three-dimensional free rotation are $E_J = BJ(J+1)$, and $(E_1-E_0) = 1.3$ meV would accordingly be obtained for the $(J=0 \leftrightarrow J=1)$ transition. In fact, with incoherent neutron scattering at 4.9 K a value of $(E_1-E_0) = 1.09$ meV was observed (Fig.6.14).

If the potential barriers are extremely high ($V_M \gg B$), each molecule will perform rotational oscillations (librations) in its potential well, which for small

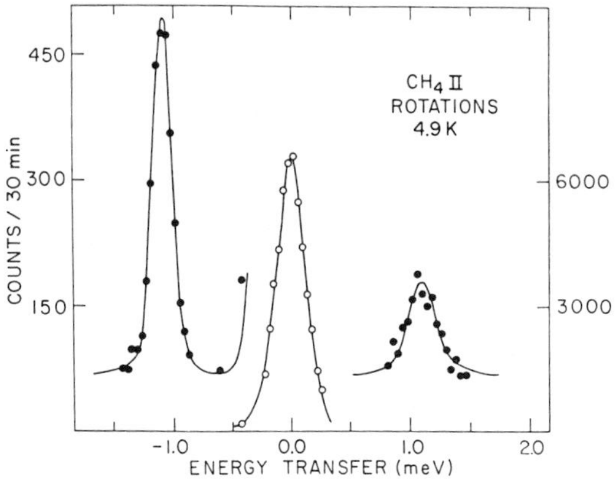

Fig.6.14. Almost free rotation of some of the molecules in Phase II of CH_4 (at 4.9 K): the maxima observed at ±1.09 meV are attributed to the ($J = 0 \leftrightarrow J = 1$) transition, for which a value of 1.3 meV would be expected in the case of completely free rotation /4.6/ (left-hand ordinate scale). The maximum at zero energy transfer (right-hand ordinate scale) is due to incoherent, purely elastic scattering

displacements can be described as those of a harmonic oscillator if the correlation with the rotational movements of neighbouring molecules is negligible. If the barriers are less high, the wave function may extend into the adjacent valleys of the rotation potential, so that the molecule tunnels with finite probability between different minima of this potential /4.5,6/. In the energy level scheme of the oscillator this corresponds to a tunnel splitting of the libration levels, which increases with decreasing potential barrier V_M. In the case of the six molecules of methane II ordered with respect to their orientation, the barriers V_M (~300 K) are considerably greater than B (~8 K), but are still of an order of magnitude which permits a convenient measurement of the tunnel splitting of the librational ground state. In accordance with the three possible different types of symmetry of the spin function within a molecule, the ground state splits into the three tunnel levels A, T and E. For reasons of symmetry the transition $A \leftrightarrow E$ is not permitted, so that only the other two transitions are observed. Thus two lines were found for CH_4-II with neutron scattering /6.30/ at approximately 70 μeV and 140 μeV, associated with the transitions $A \leftrightarrow T$ and $T \leftrightarrow E$, respectively (Fig.6.15). For $(NH_4)_2SnCl_6$, the NH_4^+ group has a rotational potential with tetrahedral symmetry, where the barriers are larger than

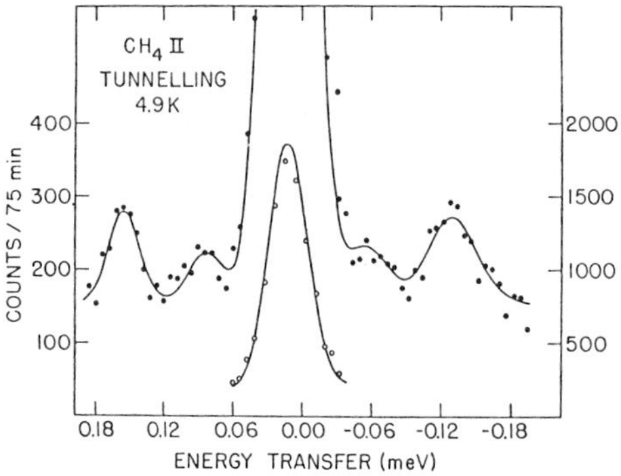

Fig.6.15. Tunnel splitting of the librational ground state of the "ordered" molecules in Phase II of CH_4. The maxima at ±70 μeV and ±140 μeV correspond to the transitions $A \leftrightarrow T$ and $T \leftrightarrow E$, respectively /6.30/ (left-hand ordinate scale). The right-hand ordinate scale applies to the elastic peak shown with reduced size

in the case of CH_4-II by roughly a factor of 2.5. The tunnel peaks observed with neutron scattering /6.31/ are at 1.5 and 3.0 μeV (Fig.6.16), and thus here the splitting is almost fifty times smaller than with methane. The tunnel splitting depends considerably on the value of the potential barrier. Clearly the investigation of the rotational tunneling effect and its variation as a function of external parameters (e.g. temperature or pressure) promises to become an extremely useful method for studying intermolecular potentials. A detailed discussion of neutron tunneling spectroscopy can be found in /6.32/.

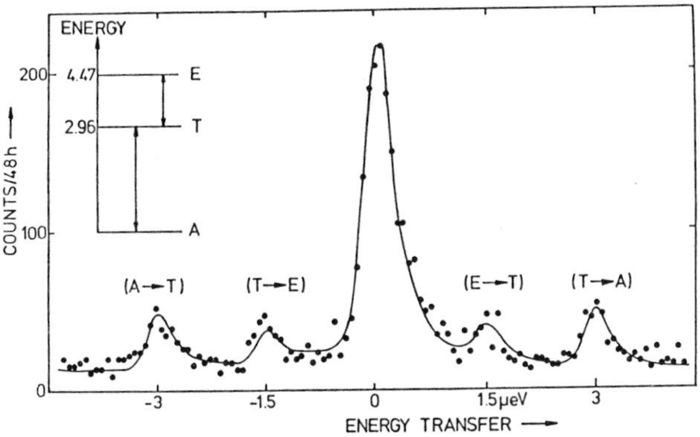

Fig.6.16. Tunnel splitting of the librational ground state of the NH_4^+ group in $(NH_4)_2SnCl_6$ /6.31/

6.7 Determination of Diffusion Parameters: Hydrogen in Metals

Neutron investigations of atomic movements related to translational diffusion have been successfully performed for more than ten years on metal-hydrogen systems. Incoherent neutron scattering permits direct measurement of the macroscopic translational diffusion coefficient D* of the hydrogen atoms by determining the quasielastic line width in the low-Q region (Sect.5.1). In addition, however, this method enables one to study the spatial and temporal aspects of atomic mobility on the microscopic scale of the order of magnitude of atomic distances, by measuring quasielastic and inelastic spectra in the range of high Q values ($|\underline{Q}| > 2/a$; a = distance between nearest-neighbour atoms). The parameters characteristic for the diffusional motion accessible to such measurements concern the spatial arrangement of the sites preferred by the hydrogen atom in the metal lattice, the average residence time per site and the frequencies the atoms contribute to the crystal lattice vibration spectrum during their residence at such a site /4.9/.

The diffusion of hydrogen in the face-centred cubic (fcc) lattice of palladium (α phase of PdH_x) demonstrates this principle. In Fig.6.17a the unit cell of this lattice (open circles) is shown with two different models of the conceivable arrangement of hydrogen positions (solid circles). According to this the hydrogen atoms in the metal lattice are either at tetrahedral or octahedral sites.

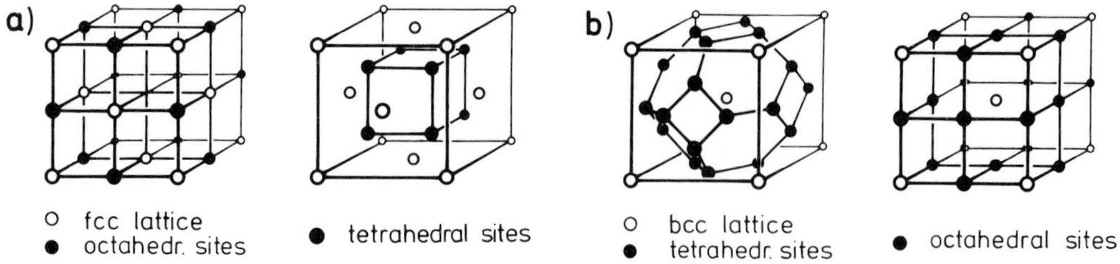

Fig.6.17. Diffusion on interstitial sites in the crystal lattice: (a) fcc lattice; (b) bcc lattice. For each lattice (o) the tetrahedral and octahedral sites for diffusing atoms (•) are shown

For high concentrations of hydrogen (> 5%) the most probable positions can be determined by neutron diffraction. At low concentrations this may be achieved with the aid of quasielastic incoherent neutron scattering because of the large incoherent scattering cross-section of H (see, e.g., /6.33/). In the case of α-PdH_x (with $0.02 \leqq x \leqq 0.04$), measurements on a polycrystalline powder /6.34/ and on a single crystal /6.35/ showed that at high temperatures the hydrogen atoms prefer the octahedral sites of the fcc Pd lattice. This was proved by comparing the quasielastic linewidths calculated from the Chudley-Elliot model (Sect.5.1) with those determined by experiment. Figure 6.18 shows this comparison for two different symmetry directions of the crystal. Whereas in the [100] direction (Fig.6.18a) tetrahedral and octahedral models give equivalent fits to the experimental data, only the octahedral model agrees with experiment in the [110] direction (Fig.6.18b). For the aver-

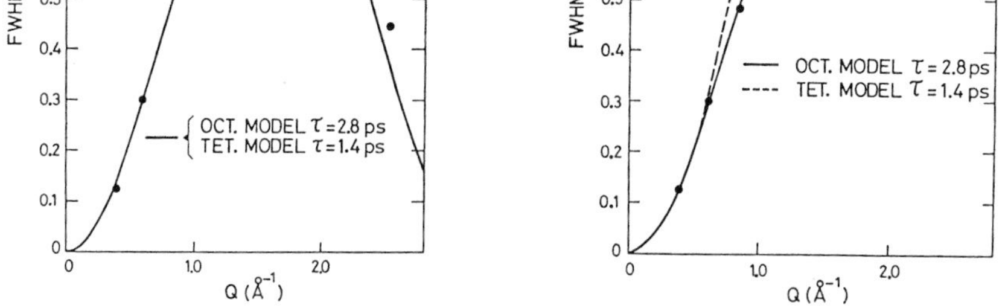

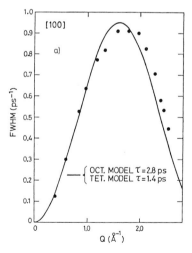

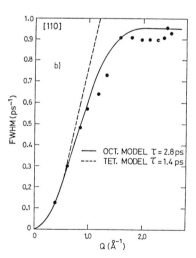

Fig.6.18. Quasielastic linewidths as a function of Q for two different crystallographic directions, (a) [100], (b) [110], in α-PdH_x. The curves represent the values calculated for two different models of the arrangement of the interstitial sites. The results shown in (b) clearly favour the octahedral-site model with an average residence time of the hydrogen atom per site of $\tau = 2.8 \cdot 10^{-12}$ s at 623 K /6.35/

age residence time of a hydrogen atom in an octahedral position, determined from the quasielastic linewidth, a value of $2.8 \cdot 10^{-12}$ s was found at 623 K.

Only a few of the metal-hydrogen systems so far studied permitted such a simple interpretation. Complex models are necessary when not one but several different types of possible sites for the hydrogen atoms occur, when the metal lattice contains impurities which attract the hydrogen or when the correlation between adjacent diffusing atoms must be taken into consideration at high hydrogen concentrations. In the first case a suitably extended Chudley-Elliott model can be applied, enabling the different average residence times associated with the various types of position to be determined.

In the presence of defined impurities of known concentration their influence on hydrogen diffusion can be studied. In the case of nitrogen impurities in niobium, which have the property of temporarily capturing hydrogen atoms and thus modifying the diffusion mechanism, it was possible to measure the capture rate and the average residence time of a hydrogen atom at the location of the defect as a function of temperature. Both quantities are governed by Arrhenius laws with activation energies which were also determined in this experiment /6.36,37/. In the incoherent scattering function explicit consideration of correlation effects to be expected at high hydrogen concentrations, e.g. similar to those observed in the case of Na /6.38/, has as far as we know not yet been attempted. This may account for the current inability to interpret such measurements satisfactorily in the range of large Q /6.39/. However, the theoretical problems disappear if one goes to the limit of small Q values, where the macroscopic diffusion coefficient D* can be measured directly (see, e.g., /6.40/) and its concentration dependence determined (see, e.g., /6.39/). This is based on the fact that at small Q the diffusion equation (5.3) is valid, independently of the local geometry of the diffusive motion and of the concentrations of impurities and diffusing atoms. It should be emphasised that the neutron scattering method, which at present permits measurements of D* in the order of magnitude from 10^{-5} cm^2 s^{-1} to 10^{-7} cm^2 s^{-1}, does not require any macroscopic concentration gradient. The diffusion mechanism can therefore be investigated at equilibrium. On the other hand, D* can also be determined in the presence of a concentration gradient, namely in real-time experiments, for instance by neutron diffraction (Chap.7) or by neutron radiography /6.41/.

7. Application of Neutron Scattering to Structural and Kinetic Problems

7.1 Kinetics of Structural Transformations

What we know about structural transformations in solids in connection with heterogeneous chemical reactions or crystallisation processes is based largely on analysis of stable or metastable compounds, inasmuch as crystallographic investigations are concerned. Kinetic processes are primarily dealt with phenomenologically. Here real-time experiments, in which the time variation of the diffraction diagram is studied, provide the possibility of developing microscopic models. The construction of efficient position-sensitive detector (PSD) systems was an essential condition for this /7.1/. The following complementary techniques may be mentioned:

1) *real-time small-angle scattering* for the investigation of structural transformations occurring in amorphous or poorly crystallised solids;
2) *real-time neutron diffraction* for all processes where a high degree of ordering is maintained, as for example in topotactic reactions.

7.1.1 Real-Time Small-Angle Scattering

Experiments on the crystallisation of glasses are particularly suitable for explaining 1). Crystallisation is often carried out in two phases in order to achieve certain material properties. First the samples are annealed below the crystallisation temperature (nucleation phase). Here very small crystalline nuclei are initially formed, which grow into nuclei of the critical size, given a sufficiently long annealing time. This initial treatment determines the way in which the growth phase takes place at higher temperatures. If there are sufficient nuclei of the critical size already available from the nucleation phase, the nuclei continue to grow immediately. However, if the majority of the nuclei have not yet reached the critical size, many of them dissolve again and an induction phase occurs, as a sufficient number of nuclei must first be formed. Such processes are well-known for decomposition reactions /7.2/.

Small-angle scattering experiments during crystallisation give information on the time variation of the radius of gyration, and thus of the particle size (Sect.6.3); they therefore permit conclusions to be drawn about the state of the glass before crystallisation. An example is a study of the crystallisation of a cordierite glass of composition $2MgO \cdot 2Al_2O_3 \cdot 5SiO_2$, containing 10% TiO_2 as nucleating agent /7.3/. This glass is particularly suitable for small-angle neutron scattering, as the crystallites precipitating from an aluminium-titanate mixed crystal give a good contrast relative to the matrix because of the negative scattering length of the titanium atom.

Figure 7.1 shows the change with time of $I(Q)$ in the growth phase (835°C) for a glass annealed for 10 hours at 720°C. An increase in intensity and the formation of a pronounced maximum at $Q = 7 \cdot 10^{-3}$ Å^{-1} are observed. The maximum is attributed to an interference effect, due to short-range ordering of the precipitating crystallites. It yields a value of ≅ 900 Å as a rough estimate of the average distance between crystallites $\Delta(= 2\pi/Q)$. This corresponds to a particle density ($\cong \Delta^{-3}$) of

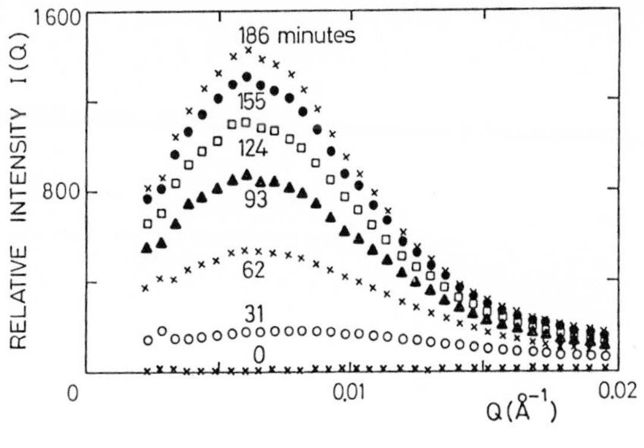

Fig.7.1. The crystallisation of a cordierite glass leads to an increase in the small-angle scattering intensity I(Q). Furthermore an interference maximum develops at $Q = 7\cdot10^{-3}$ Å^{-1}, indicating partial ordering of the precipitation /7.3/

$\cong 10^{15}$ cm^{-3}. If the actual small-angle scattering is separated, R_G can be determined by extrapolation in the Guinier region [cf. (6.3)]. Assuming spherical particles, we have for the particle diameter D /6.11/:

$$D = 1.29\ R_G \quad , \tag{7.1}$$

D increasing from ≅ 33 Å after 2 min to ≅ 350 Å after 180 min. This characterisation of the precipitation permits conclusions to be drawn about the nucleation phase and thus contributes to a better understanding of the material properties. This may be demonstrated with the aid of Fig.7.2, where I(Q) is shown for glasses with different initial heat treatments. Doubling the annealing time to 20 h leads to more nuclei of the critical size, as Δ ≅ 350 Å (particle density $\cong 2\cdot10^{16}$ cm^{-3}). The increased particle density corresponds to a lower rate of growth for the individual crystallites. Thus D reaches 110 Å only after 100 min. If the heat treatment is omitted, no interference maximum is observed, which indicates an irregular precipitation distribution. Here D ≅ 500 Å after 100 min. With longer times the irregular form of the precipitation no longer permits a precise determination of D.

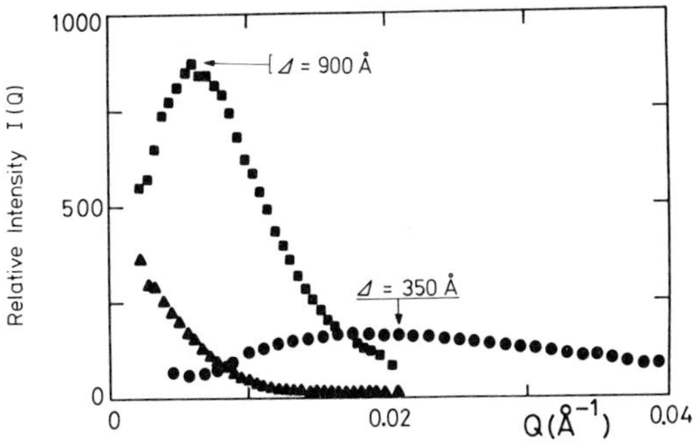

Fig.7.2. Small-angle scattering intensity of cordierite glass after 90 min heat treatment at 835°C for different annealing times. The annealing time has an effect on the position of the interference maximum. Long annealing time of the cordierite glass results in fine-grained precipitation. The mean precipitation distance (Δ) (of nearest-neighbor crystallites) was determined for various annealing times /7.3/. (■) annealed at 720°C for 12 hours; (●) annealed at 720°C for 20 hours; (▲) not annealed

7.1.2 Real-Time Neutron Diffraction

The time variation of Bragg reflections can be correlated with the formation of crystalline compounds. Model calculations concerning structure factors give information on the crystal structure. In practice, however, a lower resolution than with classical structure analysis is achieved, as the reduction of the measuring time per spectrum also limits the number of statistically significant reflections. Furthermore, the farther one goes away from thermodynamic equilibrium, the sooner can imperfections occur which, like a Debye-Waller factor, tend to weaken reflections especially at large diffraction angles.

Real-time diffraction experiments are advantageous if a quick survey of a sequence of compounds is required, such as may occur in a solid-state reaction. Here a wide field opens up for in situ investigation of compounds which can be isolated – if at all – in the metastable state only.

Topotactic reactions, in which structural features of the initial lattice and thus to some extent crystallographic order are maintained, are obvious candidates for real-time neutron diffraction experiments. Typical examples are the intercalation reactions of polar molecules in layered compounds, e.g. transition metal dichalcogenides MeX_2 (Me = Nb, Ta, Ti; X = S, Se) /7.4,5/. During such reactions a sequence of compounds can be formed.

Thus in the cathodic reduction of TaS_2 in K_2SO_4/D_2O solution

$$TaS_2 + xK^+ + yD_2O + xe^- \rightarrow K_x(D_2O)_yTaS_2 \quad , \tag{7.2}$$

several steps are observed in the potential /7.6/, which indicate the occurrence of a series of compounds (Fig.7.3). A real-time neutron diffraction experiment on a powder electrode reacting under galvanostatic conditions was carried out with a time scale of 15 min measuring time per spectrum. The charge transfer per formula unit TaS_2 and per hour amounted to 0.01 e^-. Figure 7.4 shows that the 002 reflection of TaS_2 diminishes, and that two new compounds P_1 and P_2 with characteristic reflections are formed successively. In fact the reaction is still more complicated,

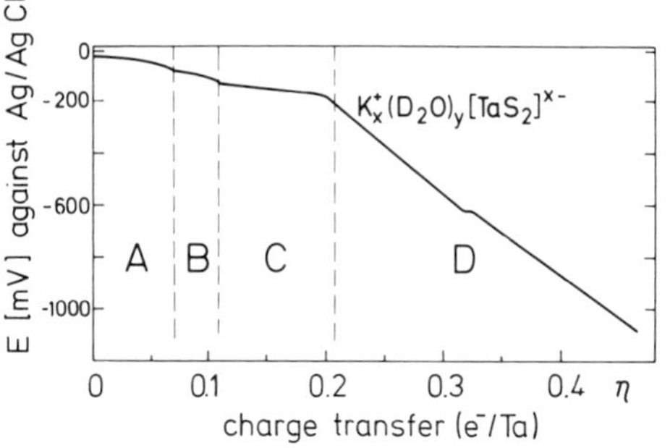

Fig.7.3. During the cathodic reduction of a 2H-TaS_2 powder electrode in a D_2O/K_2SO_4 solution several potential steps are observed, indicating the formation of several different compounds in regions A, B, C and D /7.6/

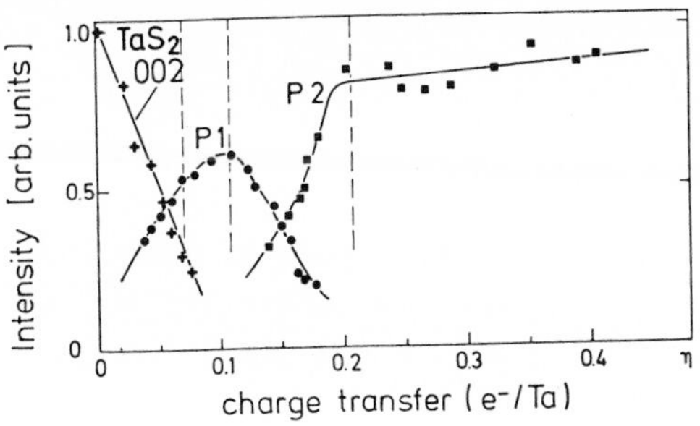

Fig.7.4. The compounds assumed in order to explain Fig.7.3 can be detected by means of two reflections, P_1 and P_2, of these compounds. The change in the intensities of these reflections and of the 002 reflection of the original lattice are correlated with the potential steps in Fig.7.3 /7.6/

as the position of the P_1 reflection changes discontinuously (Fig.7.5), which indicates two compounds whose reflections cannot be distinguished. From the positions of the reflections and the known dimensions of the basic lattice, a model can be developed for the compounds occurring in the region A → C. According to this "higher stages" initially occur, in which a $\{K_x(D_2O)_y\}^{x+}$ layer is inserted between groups of n host lattice layers (Fig.7.6). For $K_x(D_2O)_yTaS_2$ a series n = 3→2→1 was found.

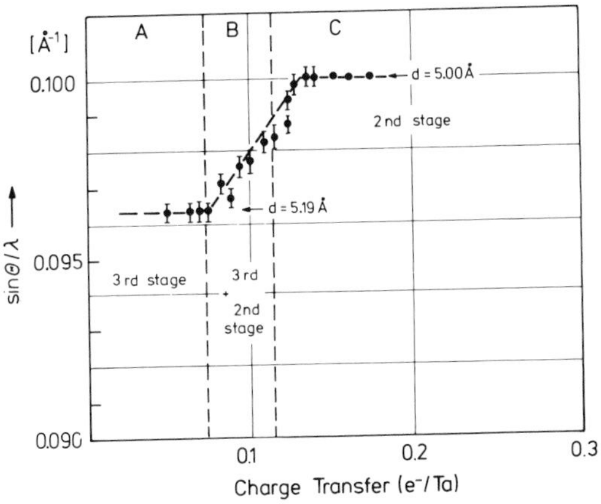

◄Fig.7.5. Two compounds in fact correspond to the P_1 reflection in Fig.7.4, as the position of this reflection changes in steps. The reflections of the two compounds are too close together to be separated /7.6/

▼Fig.7.6. A model of the compounds occurring during the formation of $K_x(D_2O)_yTaS_2$ based only on the change in the layer distance and the displacement of layers due to the trigonally prismatic coordination of the intercalated particles. The stacking order of the sulphur layers is indicated by A, B, C, and that of the metal layers by b, c /7.6/

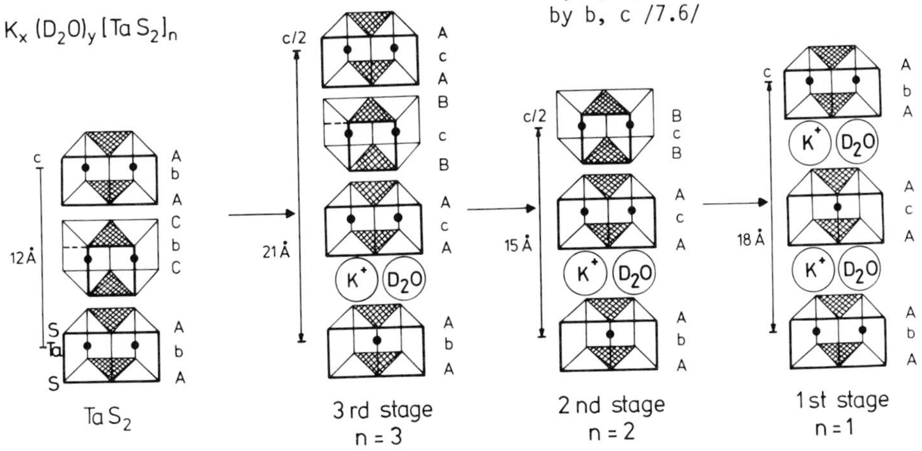

Characteristic line broadening of the 00ℓ reflections for the compounds with n = 3 and n = 2 can be explained by a perturbation in the sequence of layers, due to a random mixture of two layer packs (e.g. n = 3 with n = 2) /7.7,8/. With time resolutions of a few minutes per spectrum this disorder appears "frozen in". A higher time resolution and thus possibly experimental tests of model ideas on the dynamics of the structural transition will be possible when still better use can be made of the available neutron flux at the Grenoble High Flux Reactor with the aid of improved instruments. A new generation of PSDs under construction will permit a reduction of the measuring time per spectrum by a factor of 20 to 50 /7.8/.

The upper limit of Region B (Fig.7.5) determined by diffraction does not coincide with the step in the potential between B and C in Fig.7.3. Possibly two compounds are formed successively, which differ only slightly in the potential /7.8/. Model calculations for the system $Na_x(D_2O)_yNbS_2$, where similar compounds were found in the third-stage region /7.9/, indicate an incommensurate → commensurate structural transition. In the interlayer space only trigonally prismatic sites (TP sites) with a distance of 3.3 Å are available for the intercalated particles. In crystalline hydrates the distance between Na and H_2O is shorter by ≅ 0.4 Å. This suggests that at low cation densities the particles are displaced from the TP sites and form an incommensurate structure. Only at somewhat higher cation densities does a commensurate structure form, in which the particles occupy the TP centres. Commensurate and incommensurate structures are discussed in more detail in Sect.7.4.1.

7.2 Dynamics of Chemical Equilibria

An interesting field of application of the incoherent neutron scattering methods described in Sect.6.5 is the measurement of molecular reaction rates in dynamic equilibria. The problem here is to identify contributions in the measured scattering function $S_{inc}(\underline{Q},\omega)$ resulting from those types of atomic movement associated directly with the reaction studied. Thus spatial and temporal aspects of the local atomic arrangement are discovered (temporary specific molecular conformations and distributions of intermolecular distances, average residence times of an atom as part of one reaction partner or another), which are essential for the occurrence of a reaction. The more different atoms participate in the reaction and the more different types of movement occur in the system studied, the more difficult of course it is to isolate the desired effect. It is therefore a priori advantageous if the scattered intensity associated with the reaction to be studied is dominated by the contributions of a single type of atom. This is generally the case where hydrogen atoms are involved.

Studying the hydrogen exchange mechanism in the dissociation equilibrium of aqueous trifluorine acetic acid solution /7.10/

$$CF_3COOH + H_2O \rightleftarrows CF_3COO^- + H_3O^+ \qquad (7.3)$$

indicates how complex such an experiment is to interpret. The desired effect is on the one hand masked by the translational and rotational diffusion of the three different hydrogen-containing species. On the other hand, these types of motion should not be considered separately from the dissociation reaction, as they are necessarily strictly coupled to it because of the consecutive occurrence of the various movements – an important point which was not considered in /7.10/. In such a case it is advisable to combine different experimental techniques to reduce the number of unknown parameters in the scattering function $S_{inc}(\underline{Q},\omega)$ which completely describes the system. Valuable information can also be obtained from the study of simpler model systems, in which the various dynamic phenomena important for the chemical reaction can be observed in isolation. Such phenomena as translational and rotational diffusion, molecular conformational movements, inter- and intramolecular vibrations and exchange of atoms, ions or molecular groups between reaction partners can also be identified individually if their respective contributions to the scattered intensity depend in different ways on the experimental conditions chosen (temperature, pressure, state of aggregation, $(\underline{Q},\omega)$ range, experimental resolution).

As a relatively simple example of the investigation of an exchange mechanism coupled with other types of movement let us consider the dynamic equilibrium between the various rotational isomers in solid succinonitrile $[N{\equiv}C\text{-}(CH_2)_2\text{-}C{\equiv}N]$. This molecule takes on three rotationally isomeric conformations in the "plastic" phase of the crystal (233 K $\leq T \leq$ 331 K), whose average life times (τ_g', τ_g and $\tau_{\bar{g}}$, respectively) were determined using incoherent neutron scattering. At room temperature 22% of the molecules (fraction C_t = 0.22) have the trans form (t) and 39% each have one of the two gauche forms (g or $\bar{g}$, respectively). The different conformations are shown in Fig.7.7 by a simplified model of the molecule, whose radii and

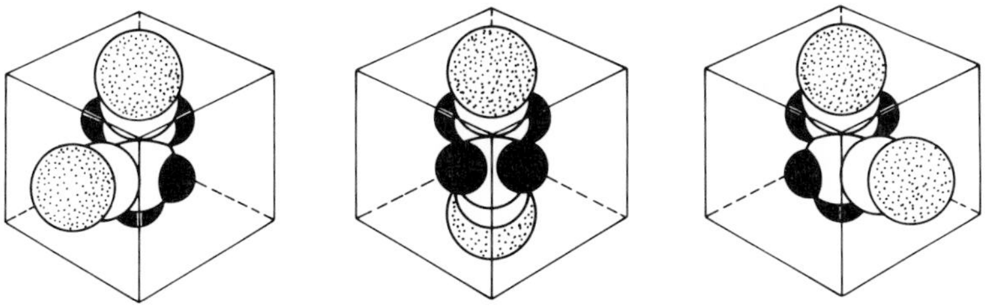

Fig.7.7. Schematic representation of the three different conformations occurring in the plastic phase of succinonitrile $[N{\equiv}C\text{-}(CH_2)_2\text{-}C{\equiv}N]$ (t = trans; g,$\bar{g}$ = gauche) relative to the cubic lattice; hydrogen atoms are black /7.11/

perspective are not drawn quite correctly in order to improve clarity. The transitions described by the equilibrium

$$g \underset{(2\tau_g')^{-1}}{\overset{(\tau_g)^{-1}}{\rightleftarrows}} t \underset{(\tau_g)^{-1}}{\overset{(2\tau_g')^{-1}}{\rightleftarrows}} \bar{g} \tag{7.4}$$

with the transition rates $1/\tau_g$ and $1/(2\tau_g')$ between the various isomers $[\tau_g' = \tau_g C_t/(1-C_t)]$ are effected by ±120° rotations of the $-CH_2-C{\equiv}N$ groups about the central C-C bond of the molecule. These rotational motions cause an energy broadening of part of the (originally elastic) neutron spectrum, where the width of this quasielastic spectrum is determined by the transition rates $[1/\tau_g$ and $1/2\tau_g'$, respectively]. The latter can therefore be determined directly by experiment. However, the problem is complicated by the fact that in addition to these rotational motions another type of movement occurs which can also contribute to the quasielastic scattering. This is 90° reorientations of the molecule about its long axis, which can take place at a rate of $1/\tau_R$ whenever the molecule has taken on its trans conformation. Measuring the EISF makes it possible to check in such a case whether the quasielastic scattering seen at a given energy resolution $\Delta\hbar\omega$ of the experiment is caused by one, by the other or by both types of movement. If the former were correct, one of the two curves in Fig.7.8 marked A (reorientation of the t molecule) and B (rotational isomerisation) would be expected to be observed for succinonitrile. In fact, however, the result of the measurement corresponds to curve C, which shows the correctness of the third alternative /7.11/. The correlation times of the various types of movement can be determined from the ($\underline{Q}$-dependent) line shape of the quasielastic spectrum. For succinonitrile at 302 K the values $\tau_R = (2.7\pm0.8)\cdot10^{-11}$ s, $\tau_g = (4.4\pm0.4)\cdot10^{-11}$ s and $\tau_g' = (1.3\pm0.2)\cdot10^{-11}$s

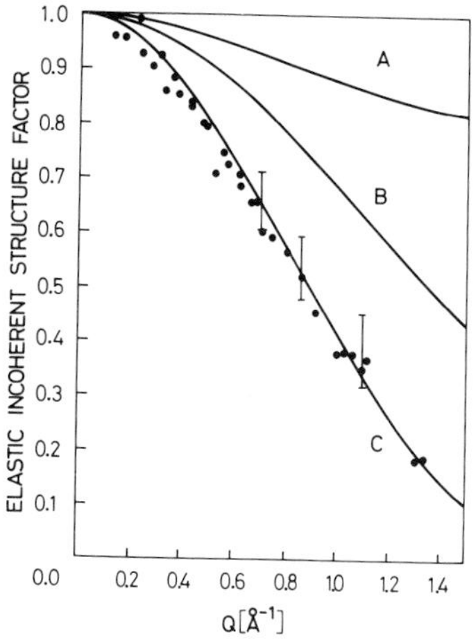

Fig.7.8. Comparison of the EISF found experimentally /7.11/ with three different models: A: only 90° rotational jumps of the t molecules about the fourfold axes of the lattice; B: only the movement corresponding to the conformation change, i.e. to the isomerisation reaction $g \leftrightarrow t \leftrightarrow \bar{g}$; C: both types of movement are allowed simultaneously

were found /7.12,13/. Although the different diffusive rotational and isomeric movements discussed here occur with comparable rates, it was possible to identify their individual contributions to the scattering function. The essential condition for this was to take into account the characteristic geometries (EISF) for the different types of motion and the strict coupling between rotational and isomeric movements.

7.3 Fast Ion Transport: Superionic Conductors

Solid electrolytes with high ionic conductivity have attracted the attention of numerous research laboratories in recent years (see, e.g., /7.14/) because of their possible applications (see, e.g., /7.15/). Similarly to metal-hydrogen systems (Sect.6.7), the neutron scattering technique can also be used with solid electrolytes to measure macroscopic diffusion coefficients and to obtain spectral information needed as a basis for developing microscopic models of ion motion. The self-diffusion of Na investigated in detail /6.38,7.16/ suggests the possibility of applying this method to superionic conductors containing Li or Na. However, with many superionic conductors the method must be somewhat different since the mobile ions mainly scatter coherently. The pair correlation of the mobile ions must therefore be taken into account.

The diffusional motion of Ag^+ ions in the body-centred cubic (bcc) iodine sublattice of α-AgI may be considered here as an example. Although the diffusion of silver ions is subject to spatial limitations due to the existence of the fixed iodine lattice, near 250°C these ions have for instance a mobility of the same order of magnitude as water molecules in water at room temperature. In Fig.6.17b the unit cell of the bcc lattice (open circles) is shown for two different models, distinguished by the arrangement of the positions preferred by the mobile ions (solid circles). By analogy with the case of the fcc lattice already mentioned (Sect.6.7), these are called tetrahedral sites (Fig.6.17b, picture on left) or octahedral sites (picture on right). (Remark: in the case of the fcc lattice (Fig.6.17a), this is a genuine tetrahedral or octahedral coordination, respectively, whereas in the case of the bcc lattice (Fig.6.17b), the coordination tetrahedra and octahedra are distorted.) Neutron diffraction experiments /7.17-19/ have made an essential contribution to settling this discussion in favor of the arrangement of Ag^+ ions in tetrahedral positions. Figure 7.9 shows the section through the Fourier synthesis of the Ag^+ ion probability density distribution /7.19/ in the (100) plane of the lattice (cube face of the cubic unit cell).

The four density maxima situated in the tetrahedral positions are clearly distinguishable. They correspond to the minima of the time-averaged potential of the

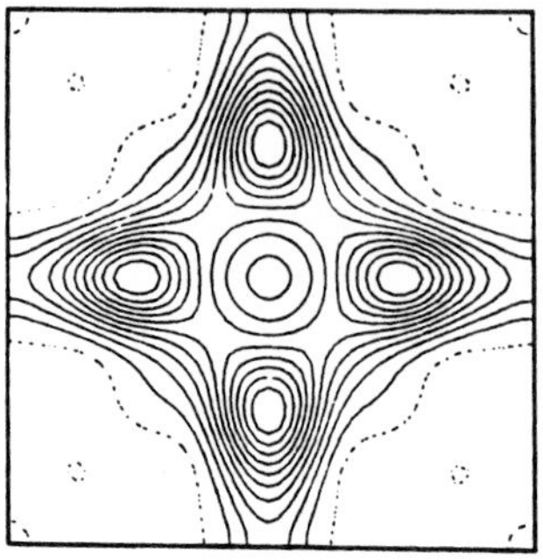

Fig.7.9. Section through the Fourier synthesis of the Ag^+-ion probability density distribution /7.19/ in the cube face of the cubic unit cell of α-AgI: density maxima are found in the four tetrahedral positions of this face

Ag^+ ions. It is clear that the fast diffusional motion of the ions, which is closely connected with the high ionic conductivity, is based on the ion transfer between these minima. An investigation of this phenomenon with quasielastic neutron scattering showed that the movement of the silver ions can in the first approximation be broken down into two components, the time constants of which differ from one another by an order of magnitude /7.20/. Whereas for long times ($t > 10^{-11}$ s) the usual diffusion equation (with a diffusion coefficient of $D^* = 2.1 \cdot 10^{-5}\ cm^2\ s^{-1}$ at 250°C) applies, for shorter periods ($t \cong 10^{-12}$ s) the superposition of a faster, spatially restricted and also diffusive movement must be taken into account. This (temporary) spatial limitation of the diffusing silver ion is caused in part by the regular arrangement of the bcc iodine sublattice. In addition to this, the Coulomb interaction with adjacent diffusing ions[10] may for short periods ($t \cong 5 \cdot 10^{-12}$ s) block possible ion transport channels and thus delay the displacement of the individual ion from its immediate vicinity /7.21/. A similar effect also seems to have been observed in "molecular dynamics" computer experiments /7.22/. It was found regarding the diffusional motion of the silver ion from one tetrahedral position to another, that jumps back to the last site occupied occurred twice as frequently as jumps to a different position. In the neutron spectrum this effect appears as a broad, quasielastic contribution under that narrower quasielastic line due to the usual translational diffusion (Fig.7.10).

Very similar neutron spectra have also been observed with oxides, such as La_2O_3 /7.23/, which have superionic crystalline phases at temperatures above 2000°C. Under these extreme conditions, neutron scattering is at present the only possible way to investigate directly the microscopic structure and dynamics of such phases.

Because of the coherent contribution to the intensity, the measured scattering function in the examples just mentioned cannot be interpreted so simply by models

[10] In contrast to the low hydrogen concentration in the example of α-PdH_x given, the high concentration of silver ions in α-AgI is necessarily determined by the stoichiometry of the compound.

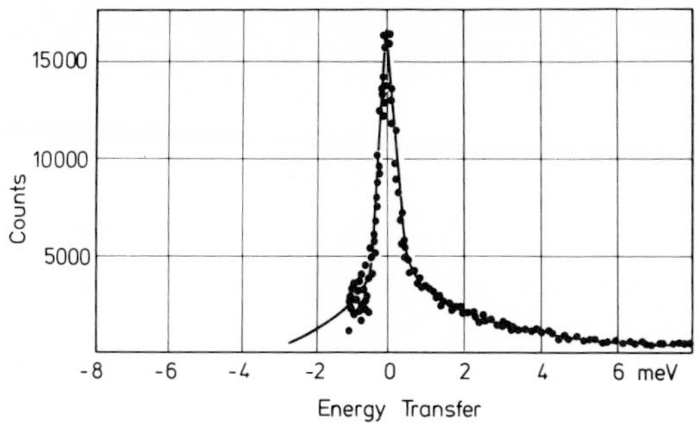

Fig.7.10. Quasielastic neutron spectrum of α-AgI at 250°C /7.20/. A clearly distinguishable narrow component (translational diffusion) is superimposed on a much broader component ("local", i.e. temporarily spatially restricted, diffusional movement)

for $S_{inc}(\underline{Q},\omega)$ as in the case of metal-hydrogen systems. Precise calculation of the coherent part necessitates explicit consideration of the time-dependent pair-correlation of neighboring diffusion ions. This problem has not yet been exactly solved, although a successfully applied approximation exists whereby the coherent scattering function is represented with the aid of a model for $S_{inc}(\underline{Q},\omega)$ and using the measured "diffuse" structure factor $S(\underline{Q})$. Even in the complete absence of incoherent scattering this phenomenological method permits an approximate determination of the diffusion coefficient D* /4.9;7.20/.

7.4 External Surfaces

Neutron scattering is best suited to study the structure and dynamics of physisorbed and chemisorbed adsorbates in certain areas due to limitations of the usual experimental methods. Thus low-energy electron diffraction experiments (LEED) can be carried out only in the pressure region $p < 10^{-4}$ torr. Heterogeneous catalysts, however, operate in a considerably higher pressure range. In addition, electron scattering techniques may lead to damage in the film structure and to desorption. It should also be mentioned that the selection rules existing in optical spectroscopy do not apply in incoherent inelastic neutron spectroscopy (IINS). The low scattering cross-sections, which are about 10^8 times smaller in comparison with those of slow electrons, result in an extremely weak signal from the absorbate layer. Pertinent neutron scattering experiments with the existing neutron sources can therefore be carried out only on powder substrates with specific surface areas of more than 2 m^2/g.

7.4.1 Structure of Physisorbed Adsorbates

The word physisorption indicates the existence of weak van der Waals bonding forces between substrate and adsorbate. Graphite is a particularly suitable substrate for neutron scattering experiments, because it is relatively transparent to neutrons and can be prepared with a large specific area. Grafoil, a frequently used variant of graphite, consists of microcrystallites with a habit in which the hexagonal (001) face is particularly prominent. Adsorption data can therefore be interpreted for this surface as a first approximation. Such thermodynamic measurements lead to a phase diagram of the adsorbate layer /7.24/, and are the starting point for neutron scattering experiments. The latter in addition provide information on:

(1) the two-dimensional (2D) unit cell of the adsorbate from the positions of the reflections;

(2) the structure of the adsorbate (e.g. orientation of molecules relative to the surface) by means of model calculations, based on the intensities observed.

As an example, Fig.7.11a shows the diffraction pattern of a single layer of the isotope ^{36}Ar on grafoil /7.25/ (^{36}Ar has a coherent scattering cross-section greater by a factor of 150 in comparison with natural Ar). The asymmetric reflection profile characteristic of a highly textured powder of two-dimensional crystallites can be described by a scattering formula developed by WARREN /7.27/. Every reflection yields a scattering angle, the integrated relative intensity and the size of the ordered region ("coherence" or correlation length).

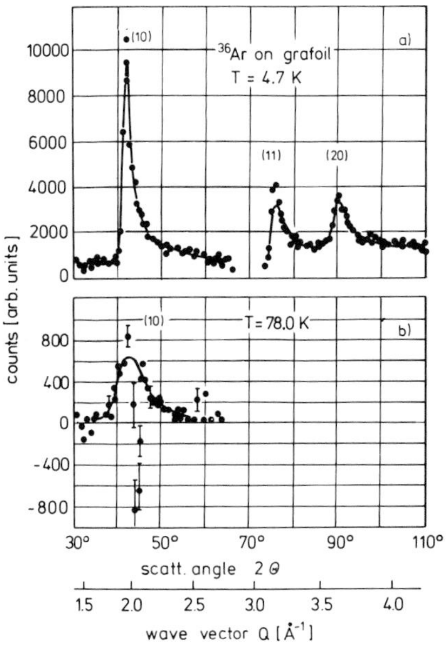

Fig.7.11. (a) The neutron diffraction pattern of a single layer of the isotope ^{36}Ar at 4.7 K physisorbed on grafoil /7.25,26/. The reflections can be indexed according to the incommensurate structure shown in Fig.7.12b. The scattering of the substrate was subtracted. (b) The same argon layer above its melting point at 78 K. The fact that only one reflection with an increased half-width is observed and that it is shifted suggests the existence of a "commensurate liquid" /7.26/

The reflections shown in Fig.7.11a can be indexed according to a two-dimensional primitive unit cell (Fig.7.12b). The Ar-Ar distance of 3.86 Å corresponds approximately to that of nearest neighbors in the 2D closest-packing arrangement of solid argon. The periodicity of the surface film, however, does not correspond to that of the substrate. This is called an "incommensurate" structure. Commensurate structures

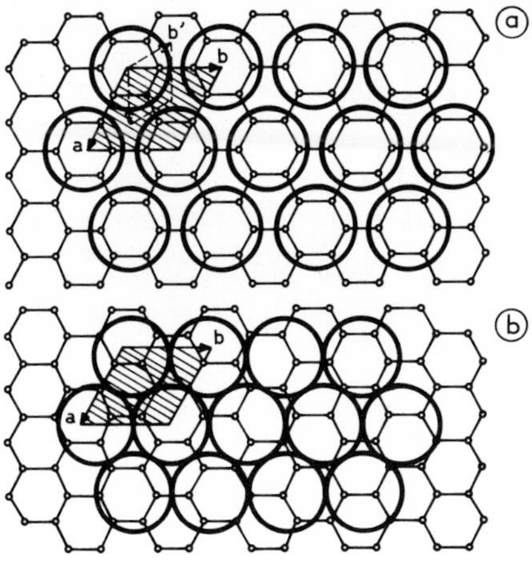

Fig.7.12. (a) Adsorption on the (001) graphite surface, after /7.25/. For the commensurate structure shown the elementary cell of the adsorbate layer (shaded) is rotated through 30° in comparison with the substrate unit cell (indicated by dashed lines), the axes are a factor $\sqrt{3}$ larger: $(\sqrt{3}\times\sqrt{3}-30°)$ structure. (b) Incommensurate structure of the adsorbate. There is no longer a simple, rational relationship between the axes of the 2D elementary cells of the adsorbate and the substrate

(Fig.7.12a) have been found for a range of simple gases such as N_2 /7.28,29/, H_2, O_2, D_2 /7.29/ and CD_4 /7.30/ on grafoil. A commensurate $(\sqrt{3}\times\sqrt{3}-30°)$ structure is frequently formed first, which changes into an incommensurate structure at a critical degree of surface coverage (Θ). The position of the 10 reflection thereby changes. This is shown in Fig.7.13 for D_2 adsorbed on grafoil /7.29/. In the region of the $(\sqrt{3}\times\sqrt{3}-30°)$ structure the position of the 10 reflection does not change. Above the limiting capacity of this structure ($\Theta > 1$) a jump is observed to the new, approximately constant position of the reflection of the incommensurate structure. This structural transition also manifests itself by a change in intensity of the 10 reflection. Thus the intensity increases linearly with the adsorption of N_2 on grafoil, until the limiting capacity of the $(\sqrt{3}\times\sqrt{3}-30°)$ structure is reached at $\Theta = 1$, Fig.7.14 /7.28/. For $\Theta > 1$ the commensurate structure is reduced, so decreasing the intensity of the 10 reflection of this structure.

Commensurate and incommensurate arrangements differ in the relative strengths of the adsorbate-substrate (A-S) and adsorbate-adsorbate (A-A) interactions. Whereas

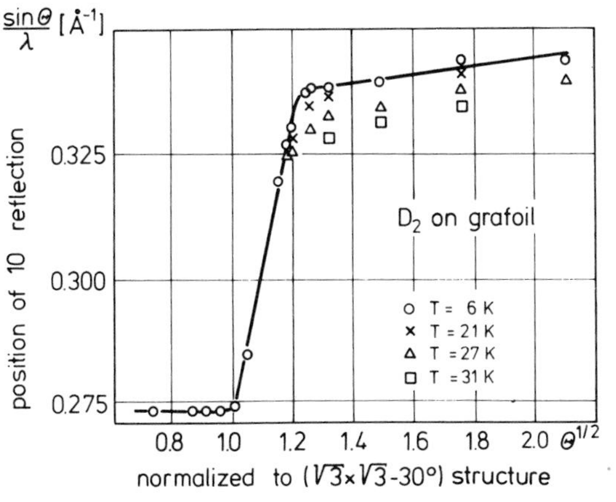

Fig.7.13. Change in the position of the 10 reflection of D_2 on grafoil as a function of the square root of the degree of surface coverage $\Theta^{1/2}$. The $(\sqrt{3}\times\sqrt{3}-30°)$ structure is present up to $\Theta = 1$, while for $\Theta > 1$ the incommensurate structure is formed /7.29/

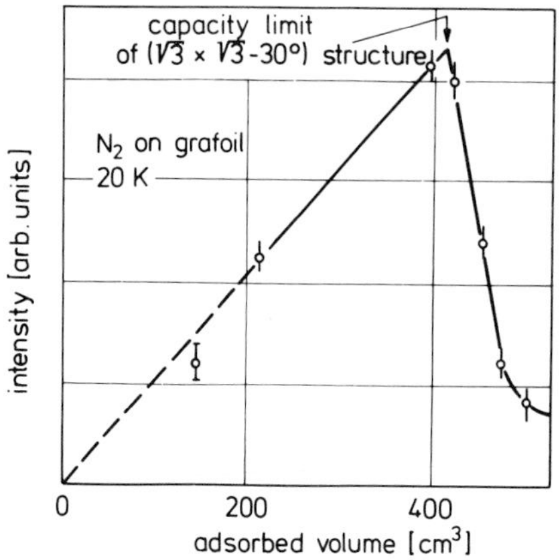

Fig.7.14. The increase in intensity of the 10 reflection of the $(\sqrt{3}\times\sqrt{3}-30°)$ structure (Fig.7.12a) for N_2 on grafoil is correlated with the adsorbed amount of N_2 /7.28/. The limiting capacity is reached when 1/3 of the carbon hexagons on the surface are occupied by N_2. Additional N_2 adsorption leads to the formation of the incommensurate structure (Fig.7.12b) whose 10 reflection has a different position (Fig.7.13). A reduction in intensity of the 10 reflection of the commensurate structure is accordingly observed /7.28/

the (A-S) interaction is dominant with commensurate structures, the (A-A) interaction is more important with incommensurate structures. The assumption of a two-dimensional solid must, however, remain an approximation, as a finite (A-S) interaction always exists. This is seen for example in the phonon spectrum of Ar on grafoil, the analysis of which showed that the mean square displacement of the argon atoms parallel to the surface, $\langle u_{\parallel}^2 \rangle$, has only 50% dynamic character /7.25/. The additional static distortion of the adsorbate layer is explained by a local adaptation of the argon atoms to the substrate potential. According to a theoretical model this is achieved by means of a cooperative displacement of the argon atoms consisting of a rotation of the adsorbate layer through a few degrees from the orientation of the commensurate structure /7.31/. LEED experiments on the Kr/graphite system at low temperatures confirm this and show krypton domains rotated through ±3.5° in comparison with the $(\sqrt{3}\times\sqrt{3}-30°)$ structure /7.32/. These domains are distinguishable by satellite reflections appearing close to the main reflection. To analyse the small distortions of the reflection profile resulting from this at the structural transition requires the construction of diffractometers with particularly high resolution. Because of the high photon flux, this is possible especially for synchrotron radiation in the X-ray range. It was thus possible to confirm experimentally the existence, predicted with the aid of refined theoretical models, of two different domain structures occurring at different Θ values /7.33/.

A precise analysis of the reflection profile is also necessary to investigate the "melting" of adsorbate layers. It is an important criterion for the presence of a liquid surface layer that the coherence length L is of the order of a few nearest-neighbor atomic distances. The "reflection" of a 2D liquid, broadened in comparison with the reflection of a 2D solid, is shown in Fig.7.11b for the example of argon on grafoil. Theoretical models of the melting process in the 2D solid start from the assumption that dislocation pairs exist, which dissociate continuously above the melting point /7.34/. This should lead to a continuous diminution of L.

Although such a diminution was observed for the Ar/grafoil system /7.25/, this interpretation is still controversial /7.35/. The melting process of commensurate structures is less disputed. According to the theory this is a 2D phase transition of a lattice gas, which may be of first or second order. Both types of transitions have been found for the Kr/graphite system with only slightly different degrees of coverage by synchrotron radiation /7.33/.

Only a small part of the present theoretical and experimental investigations concerning solid-solid and solid-liquid 2D phase transitions of adsorbates can be given here. A more detailed survey may be found in /7.36,37/.

Further structural information (e.g. the orientation of molecular adsorbates on the surface) is obtained in principle by model calculations based on the observed reflection intensities. The fact that here the kinematic scattering theory can be applied is an important advantage in comparison with LEED, where the dynamic scattering theory must be used. Apart from the superlattice reflections, the modulation due to an adsorbate layer of the intensities of substrate reflections can be quantitatively determined. This is shown in Fig.7.15 for adsorption on a substrate consisting of MX_2 layers. We consider only substrate reflections, whose scattering vectors coincide with the layer normal. If I_s and I_{s+a} are the intensities of a substrate reflection before and after adsorption, respectively, then /7.25,38/

$$I_{s+a}(Q_z) \cong \left| b_s \sum_{n=0}^{N-1} \exp(iQ_z nd) + b_a \exp[iQ_z(Nd+\delta)] \right|^2 \quad . \tag{7.5}$$

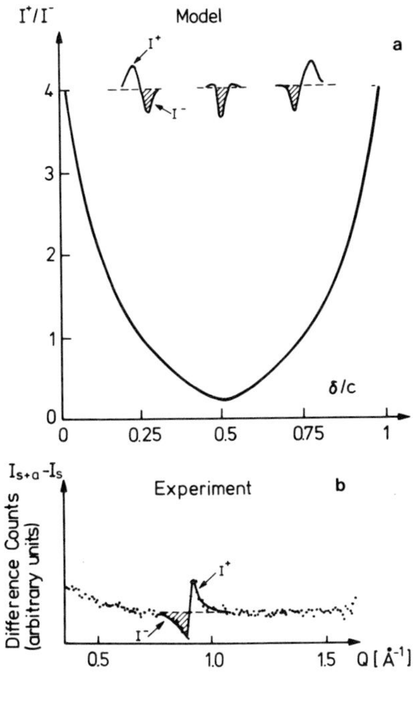

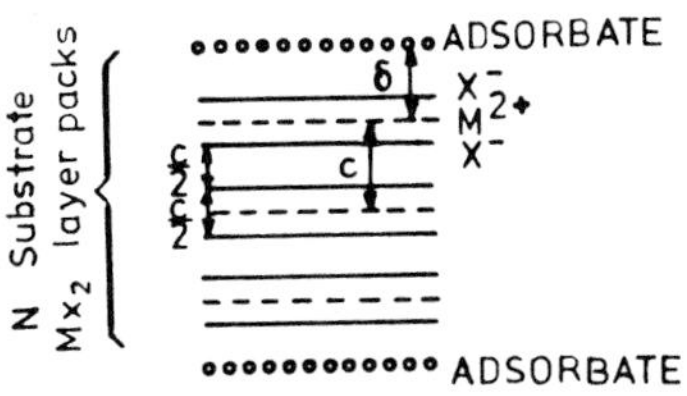

Fig.7.15. a: Model calculations concerning the modulation of the intensity of a substrate reflection due to the presence of an adsorbate layer /7.38/. The difference intensities are shown schematically for 3 different distances between adsorbate and substrate surface (δ). I^+ and I^- correspond to the amounts of integrated intensity above and below the difference spectrum zero line. The ratio I^+/I^- was calculated as a function of δ/c, where c corresponds to the distance of the metal atom layers in MX_2 layer lattices (e.g. PbI_2); c is the distance called d in (7.5). b: Example of the experimentally determined difference intensity of CD_4 on PbI_2. From $I^+/I^- \cong 1$ one obtains $\delta/c \cong 0.75$

Here b_a and b_s are the scattering lengths of the adsorbate and substrate atoms, and d is the layer repeat distance. The other quantities are defined in Fig.7.15 (where the layer interval d is indicated by the letter c).

This equation enables a simple determination of the height (δ) of an adsorbate layer above the last substrate layer. In practice the difference spectrum $I_{s+a}-I_s$ is determined, as otherwise the small intensity modulations are often difficult to distinguish. The values I^+ and I^-, whose ratio characteristically depends on δ/d, are deduced from the difference spectra (Fig.7.15). For the example of a CD_4 layer on PbI_2 a ratio $I^+/I^- \cong 1.0$ was obtained experimentally. This yields a value of $\delta/d \cong 0.75$, cf. (7.5).

Reflections with a scattering vector perpendicular to the layer normal can be analysed similarly. Here information is obtained on the position of the adsorbate in the substrate plane /7.39/. Such model calculations have been effected quantitatively with success only for spherically symmetric adsorbates. For non-spherical molecules the orientation can be qualitatively determined by the method shown in Fig.7.15. In this way it was found that n-butane molecules (C_4D_{10}) lie flat on the surface of graphite /7.40/. On the other hand, conclusions can be drawn about the formation of adsorbate layers from the change in intensity of substrate reflections. In the case of powders with preferential formation of several types of surfaces, this should permit the determination of the relative proportion of each surface. Thus from the adsorption of CD_4 on γ-Al_2O_3 it was concluded that the total surface area of the crystallites consists of ~83% (110) faces and ~17% (100) faces /7.41/. This method might become important for the characterisation of catalyst powders.

With larger molecules, larger elementary cells and more reflections are generally observed. This permits more precise information to be obtained on the structure of the adsorbate layer. Thus for deuteroethane (C_2D_6) on grafoil at 86 K ($\Theta = 0.8$), seven superlattice reflections were observed /7.42/. In the model calculations the orientation of the molecule relative to the surface was systematically modified by rotation about three orthogonal axes. A structure was obtained in which the molecules are tilted by about 24° away from the plane of the substrate and touch the surface with one deuterium atom. With the increasing number of reflections, the possibility of overlapping adjacent reflections also increases. For this reason in powder diffractometry generally the overall profile of the diffraction diagram is calculated /7.43/. This method was first used for butane (C_4D_{10}) on grafoil, as several reflections of the adsorbate overlap /7.44/. Here, too, it was found that the carbon skeleton is tilted away from the substrate plane (30° ± 5°). This example also shows that determination of the orientation from the modulation of substrate reflections merely gives an approximate picture (see above). The only aromatic molecule so far studied on grafoil is benzene /7.45/; this molecule lies flat in the 2D solid. A discussion of the 2D phase transitions and structures of further adsorbed molecules is given in /7.35/.

7.4.2 Dynamics of Physisorbed Adsorbates

Spectroscopic studies of the diffusion of hydrogen containing adsorbates with quasi-elastic neutron scattering (QNS) complement the structural investigations, which are limited to the crystalline surface phases. Results from QNS measurements of methane (CH_4) on graphite suggest the existence of three surface phases. Above the critical temperature T_k (70 K) the full-width-at-half-maximum (FWHM) of the quasielastic spectrum decreases with increasing values of Θ /7.46/. As the measurements were limited to a Q range with $Q \ll \pi/a$, where the effects of molecular rotation (with radius a) can be neglected, it was concluded that Θ is directly related to the self-diffusion coefficient D*.

This can be explained with the aid of a surface gas model, in which D* depends on the concentration of holes $(1-\Theta)$ in the adsorbate layer. Below T_k the FWHM and thus D* are independent of Θ (Fig.7.16a) /7.47/. This indicates a more strongly associated surface phase which can be characterised as liquid. In the region of the solid phase ($T < 70$ K), librational vibrations and tunnel splittings of the librational levels are observed (Fig.7.16b) /7.48/. Model calculations suggest a rota-

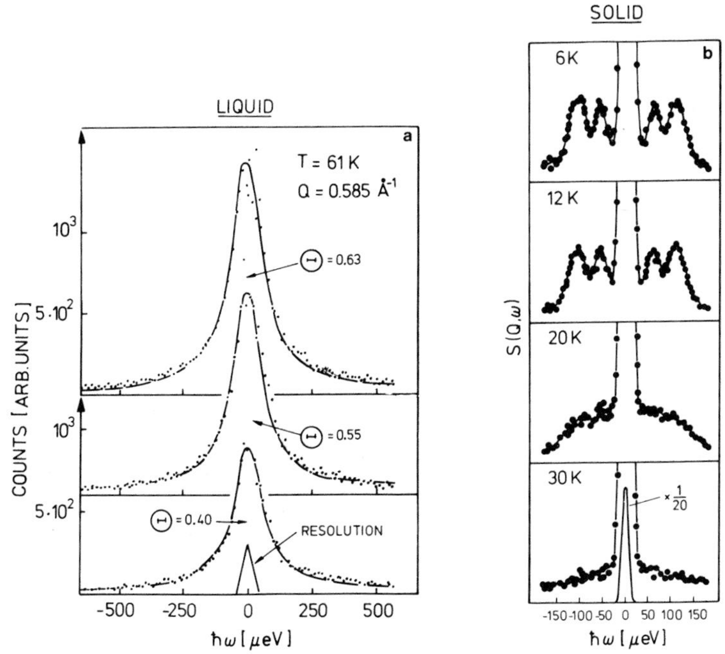

Fig.7.16a. Quasielastic neutron spectrum (QNS) of a methane layer on grafoil at 61 K /7.47/. The diffusion coefficient (D*) can be determined directly from the full-width-at-half-maximum (FWHM) (similarly to the case of three-dimensional diffusion, Sect.5.1). That the FWHM and thus D* do not change with the degree of surface coverage Θ indicates a more strongly associated liquid surface phase

Fig.7.16b. Below 30 K the movement of methane on graphite is increasingly "frozen" in. Tunneling bands then occur in the QNS spectra /7.48/. (Measurements made at $Q = 1.0$ Å^{-1})

tion of the methane molecules about a C_3 axis perpendicular to the surface. Although these are only preliminary results, this experiment appears to be of particular importance, as the heights of the potential barriers for the rotation of the methane molecules are closely related to the values of the librational level splittings and a quantitative investigation of the adsorbate-adsorbate interaction thus appears possible (cf. pure methane, Sect.6.6).

7.4.3 Chemisorbed Adsorbates

"Chemisorption" means the existence of covalent or ionic bonding between substrate and adsorbate. Only a few adsorbates, such as oxygen, are capable of chemisorption on grafoil. No other powder substrates with a large well-defined specific area are at present available. The particular advantages of neutron scattering, especially as an "in situ" method at high pressures and high temperatures, also make experiments on less well-defined surfaces interesting. Thus simply the detection of crystalline surface phases by neutron diffraction can represent important information. Reference should be made here to the underpotential deposition of metals on metal substrates /7.49/. The electrochemical data indicate the existence of ordered adsorbate structures at the phase boundary between metal and electrolyte, similar to those found at the phase boundary between metal and vacuum. The proof of the crystalline character of these surface phases by means of an in situ method has, however, not yet been successful. This should be possible with neutron diffraction, if one manages to produce substrate electrodes with a sufficiently large surface area. The feasibility in principle of such experiments has been demonstrated for the passivation of nickel powder /7.50/. However, no reflections of a crystalline surface phase have been found.

The interaction of hydrogen with metal substrates in powder form has been studied much more extensively by means of incoherent inelastic neutron spectroscopy (IINS) in the optical range of the phonon spectrum. These experiments are of importance for the understanding of the processes occurring in heterogeneous catalysis. The adsorption of water on Raney nickel may be mentioned as an example /7.51/. Three hypothetical adsorption mechanisms are:

$$\text{physisorption:}\quad H_2O \rightarrow H_2O_{ads} \quad ; \tag{7.6}$$

$$\text{chemisorption:}\quad H_2O \rightarrow H_{ads} + OH_{ads} \tag{7.7a}$$

$$H_2O \rightarrow 2H_{ads} + O_{ads} \quad . \tag{7.7b}$$

The IINS spectra of 1.5 mg H_2O/g Ni show the characteristic bands also found for hydrogen on Raney Ni (Fig.7.17a,c). Both bands were attributed to multiply bonded

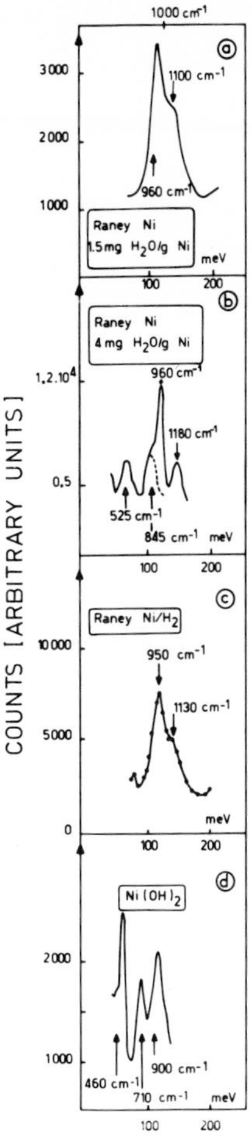

Fig.7.17. Incoherent inelastic neutron spectroscopy permits simple characterisation of the type of adsorbate. If small amounts of water are adsorbed on Raney nickel (a), a spectrum is obtained corresponding to that of H_2 on Ni (c). Chemisorption thus takes place with dissociation of H_2O. An additional physisorption of H_2O is observed for 4 mg H_2O/g Ni (b). A comparison with the $Ni(OH)_2$ spectrum (d) enables the formation of hydroxide to be excluded (data taken from /7.51/)

hydrogen. New bands occur only from 4.0 mg H_2O/g Ni upwards (Fig.7.17b). Formation of hydroxyl groups (7.7a) was excluded by comparison with the spectrum of $Ni(OH)_2$ (Fig.7.17d). A comparison with the IR spectra of crystalline hydrates leads one to suppose that the (525 cm^{-1}) band in Fig.7.17b is attributable to a "rocking" vibration of water molecules. Here an additional physisorption occurs which is superimposed on the dissociative chemisorption according to (7.7b). A summary of further IINS experiments, in particular on technologically important systems such as Pt/H_2, MoS_2/H_2, may be found in /7.52/.

7.5 Internal Surfaces

We use here the expression "internal surface" to classify solids which contain one-, two- or three-dimensional structural units and are capable of storing atoms or molecules – in the ideal case reversibly – across an interface. Such topotactical reactions have already been discussed in Sect.7.1. Especially the reactions of solids with layered structures (intercalation reactions) are of interest for fundamental and applied research. Reference should be made to recent thorough studies of transition metal dichalcogenides, MeX_2 (Me = Ti, Ta, Nb, V, Mo; X = S, Se), which are of interest as catalysts /6.5/, solar cells /7.53/, superconductors /7.4/ and fast ion conductors /7.54/. Such materials are used in fundamental research as model systems for adsorption on defect-free surfaces.

7.5.1 Structure of Intercalation Compounds

As the host lattices often show considerable imperfections (e.g. layer-stacking faults), only a few three-dimensional structural studies have been carried out on intercalation compounds. The orientation of intercalated molecules can, however, often be determined by a Fourier projection of the nuclear scattering density on the normal to the layers.

For light-atom molecules in heavy-atom host lattices neutron scattering has advantages in comparison with X-ray diffraction. In addition, however, the difference in the scattering lengths of H and D can be utilised to localise these atoms in a one-dimensional nuclear scattering density projection. This may be illustrated by the example of $NbS_2(C_5H_5N)_x$[11]. In the course of intercalation the electron concentration in the conduction band increases /7.55/. This suggests the orientation shown in Fig.7.18a I, which permits optimum interaction of the free electron pair of the nitrogen atom with the layers.

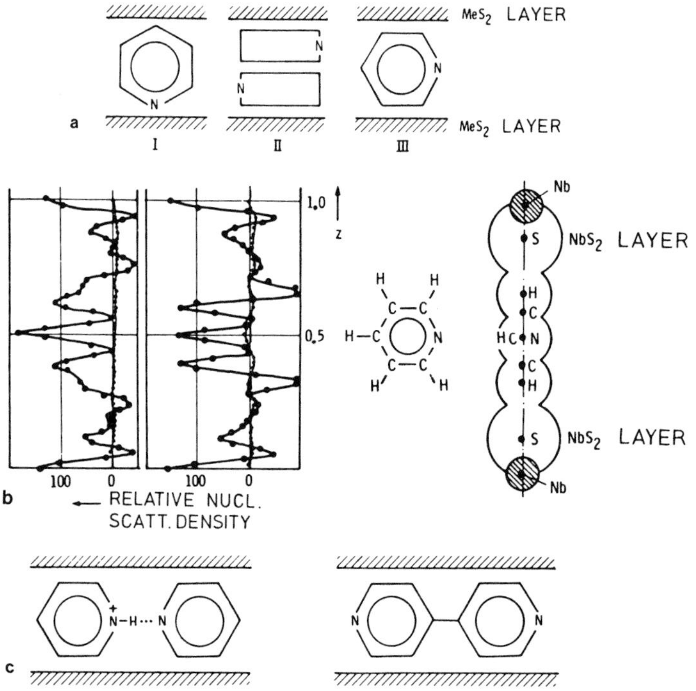

Fig.7.18. (a) Three models concerning the orientation of the pyridine molecules in $MeS_2(C_5H_5N)_{0.5}$ (Me = Nb, Ta). (b) One-dimensional Fourier projection of the nuclear scattering densities of $NbS_2(C_5D_5N)_{0.5}$ (left-hand curve) and $NbS_2(C_5H_5N)_{0.5}$ (right-hand curve) /7.56/. The H atoms next to the layers become perceptible - in contrast to the corresponding deuterium atoms - as minima in the relative nuclear scattering density. The orientation of the pyridine molecule corresponding to Model III, as well as the packing, are also shown. (c) The orientation of the molecules is explained on the basis of chemical analysis by the formation of dipyridine molecules and pyridinium cations, which bind further pyridine molecules /7.58/

[11] During intercalation a series of non-stoichiometric compounds is formed (Sect. 7.1.2). To characterise these, one usually just indicates the amount of the intercalated substance per simplest formula unit of the host lattice. In the present case 0.5 C_5H_5N molecules per formula unit NbS_2 can be intercalated at most (x = 0.5).

Model II and III shown in Fig.7.18a cannot, however, be excluded on the basis of the observed lattice expansion. In fact the experimentally determined nuclear scattering densities /7.56/ for $NbS_2(C_5H_5N)_{0.5}$ and $NbS_2(C_5D_5N)_{0.5}$ agree only with Model III (Fig.7.18b). This model is particularly unfavorable for a charge transfer from the free electron pair of nitrogen to the host lattice. This is also found in the IINS spectrum, which is only slightly modified in comparison with that of solid pyridine /7.57/. Chemical analysis leads to the most convincing explanation to date of the scattering densities, as ~40% of the intercalated molecules were recovered as 4,4' dipyridine ($C_{10}H_8N_2$) /7.58/. It is assumed that dipyridine molecules are produced according to

$$2C_5H_5N \rightarrow C_{10}H_8N_2 + 2H^+ + 2e^- \tag{7.8}$$

and that the electrons are taken up by the host lattice layers. To balance the charge, pyridinium cations and dipyridine molecules are intercalated and the former fix one pyridine molecule each by means of a hydrogen bond (Fig.7.18c). With this model the mass distribution in the Fourier projection remains unchanged.

Real-time neutron diffraction experiments on the exchange between C_5D_5N and C_5H_5N in $TaS_2(C_5H_5N)_{0.5}/C_5D_5N$ appear to confirm this model /7.59/. The change in the H/D ratio was followed in real time in the van der Waals gap. It was found that only about 50% of the molecules can be replaced, which may be explained by a lower mobility of the dipyridine molecules. Accompanying chemical reactions may be presumed with many intercalations of polar molecules in chalcogenide lattices. Thus the formation of NH_4^+ cations was proposed as a byproduct of ammonia intercalation in TaS_2 /7.60/, and $TaS_2(NH_3)_{1/3}(H_2O)_{2/3}$ was formulated as $(NH_4^+)_{1/3}(H_2O)_{2/3}[TaS_2]^{1/3-}$ /7.61/. This was confirmed with IINS spectroscopy /7.62/.

7.5.2 Spectroscopic Studies

As already mentioned in the section on surfaces, spectroscopic studies on diffusion provide important information on short-range order in low-dimensional systems. For $TaS_2 \cdot NH_3$ the possibility arose of measuring the quasielastic spectrum on single crystals over a broad $\underline{Q}$ range, and of developing a microscopic jump model, similar to those for H in metals (Sect.6.7). Figure 7.19 shows the immediate vicinity of a NH_3 molecule in $TaS_2 \cdot NH_3$ /7.63/. As the sulphur layers are directly above each other, only trigonal prismatic holes are possible, of which only 50% can be occupied in a stoichiometric compound due to the space requirement of this molecule. The compound, however, contains about 10% vacancies in the ammonia sublattice, so that diffusional jumps via vacancies (Model I, jump length 3.32 Å) or via interstitial sites (Model II, jump length 1.92 Å) appear possible. The Chudley-Elliott concept /5.2,3/ dis-

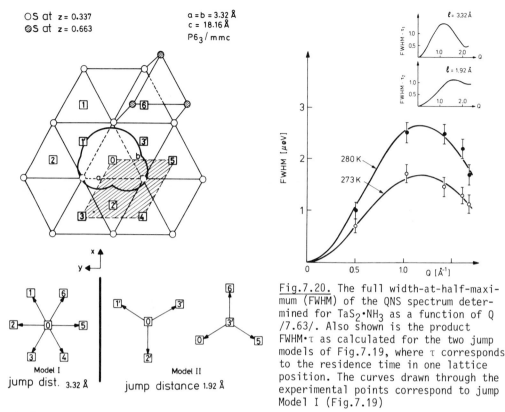

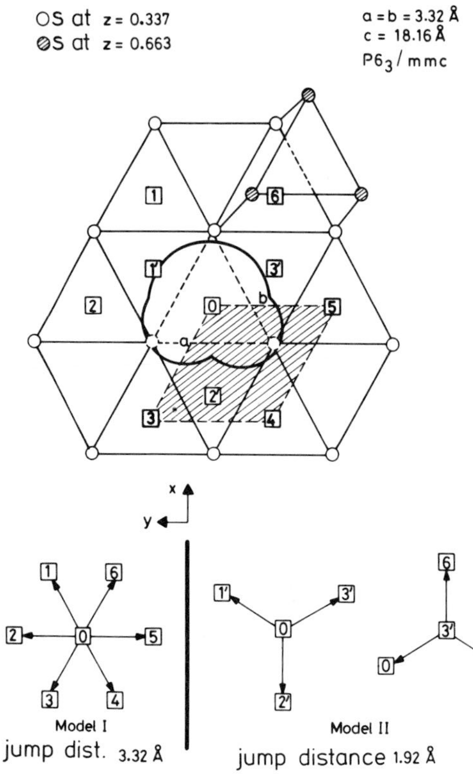

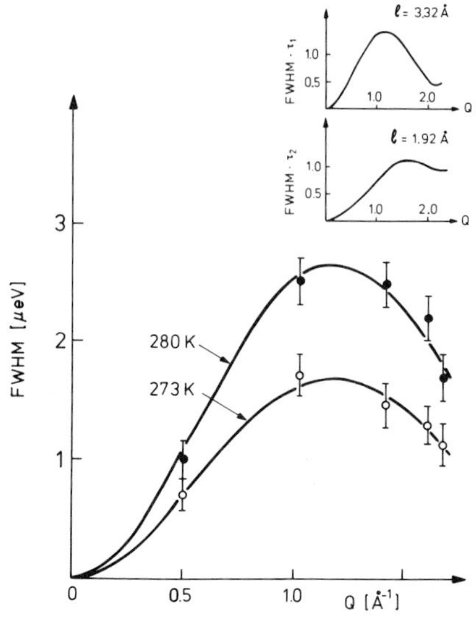

Fig.7.20. The full width-at-half-maximum (FWHM) of the QNS spectrum determined for $TaS_2 \cdot NH_3$ as a function of Q /7.63/. Also shown is the product FWHM$\cdot\tau$ as calculated for the two jump models of Fig.7.19, where τ corresponds to the residence time in one lattice position. The curves drawn through the experimental points correspond to jump Model I (Fig.7.19)

Fig.7.19. Part of a sulphur layer (o) in $TaS_2 \cdot NH_3$ /7.63/. The next layer is shown with shaded circles. An ammonia molecule has been drawn with its size based on van der Waals radii. Around the ammonia molecule there are three trigonally prismatic holes (TP) at a distance of 1.92 Å and six at a distance of 3.32 Å, leading to the jump models shown

cussed in Sect.5.1 was applied to both jump models with the residence time τ as the only parameter. Figure 7.20 shows that the values of the product FWHM$\cdot\tau$ are significantly different for the two models. The experimental data /7.63/ agree very well with Model I (diffusion via vacancies). The jump rate of $\tau^{-1} = 2.7 \cdot 10^9\ s^{-1}$ at 300 K is in very good agreement with NMR results /7.64/. Since the effect of the rotational movement on the QNS spectrum becomes increasingly important for $Q > \pi/2a$ (a being the radius of rotation), it was also taken into account. A deconvolution of the QNS spectrum is possible if the time scales of rotational and translational motions are sufficiently different. For $TaS_2 \cdot NH_3$ there were in fact indications in the QNS spectrum of a further, more rapid process, which probably corresponds to the reorientation of the molecules about an axis perpendicular to the TaS_2 layers.

The interpretation of the slowest process as translational diffusion is supported by a real-time neutron diffraction experiment on the exchange in $TaS_2 \cdot NH_3/ND_3$ /7.65/.

It was found that the change in intensity of several Bragg reflections depends on the H/D ratio in the van der Waals gap and follows a $\sqrt{t}$ law. With the known particle size a self-diffusion coefficient D* was determined from this experiment, which corresponds to the value obtained from the QNS experiments to within a factor of ~2. These measurements exclude a rate-determining proton transport as found in layered silicates by real-time neutron diffraction /7.66/.

Although the diffusion process follows the periodicity of the metal sublattice this is no proof of a direct interaction between the host lattice and the intercalate, as in the presence of NH_4^+ cations /7.60,62/ solvate molecules might diffuse between regularly arranged cations. Therefore it is first necessary to clarify whether similar jump vectors also occur with other molecules.

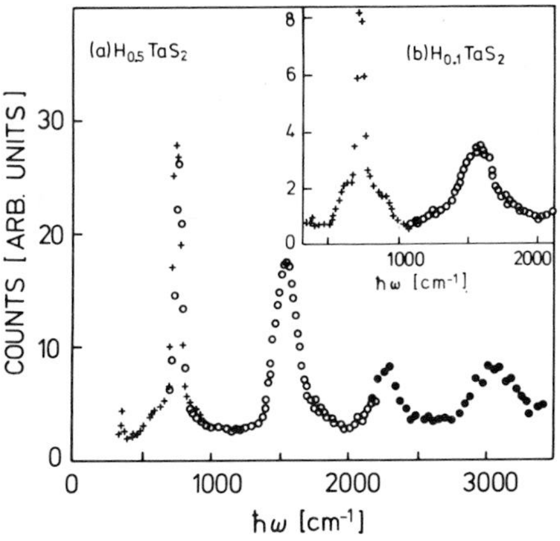

Fig.7.21. Incoherent inelastic neutron spectra (IINS) of $H_{0.5}TaS_2$ and $H_{0.1}TaS_2$ /7.67/. Three overtones are observed in addition to the fundamental frequency at 744 cm^{-1}

Spectroscopic studies in the optical range (IINS) have been carried out in particular on hydrogen-containing bronzes H_xMeO_3 (Me = Mo, W) and on chalcogenides. The spectrum of $H_{0.5}TaS_2$ (Fig.7.21) exhibits a fundamental vibration at 744 cm^{-1} and three harmonics /7.67/. It was shown that this behavior corresponds to a deformational vibration of the hydrogen atom away from the center of the three metal atoms (Sect.6.1) and in c direction /6.4/. For H_xWO_3 and H_xMoO_3 (Me-O-H)-deformation vibrations were found /7.68/. The possibility of carrying out a fast and non-destructive analysis of the bond state of the hydrogen atom complements the often tedious structural studies. Thus the MoO_3 structure consists of MoO_6 units combined to form layers. For $D_{0.36}MoO_6$ the deuterons were found localised within the layers and on oxygen atoms /7.68/. The IINS spectrum of $H_{0.34}MoO_3$ correspondingly shows a (Mo-O-H) deformational vibration at 1267 cm^{-1}. With increasing hydrogen content (x_{max} = 2.0) the deformation band diminishes and new bands associated with librational and deformational vibrations of water molecules appear as demonstrated in the correlation diagram (Fig.7.22). The hydrogen atoms are probably in between the layers, which would explain the high proton mobility found with NMR methods /7.69/.

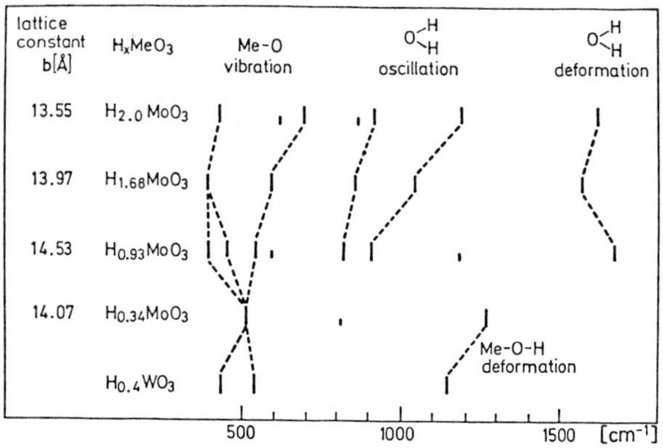

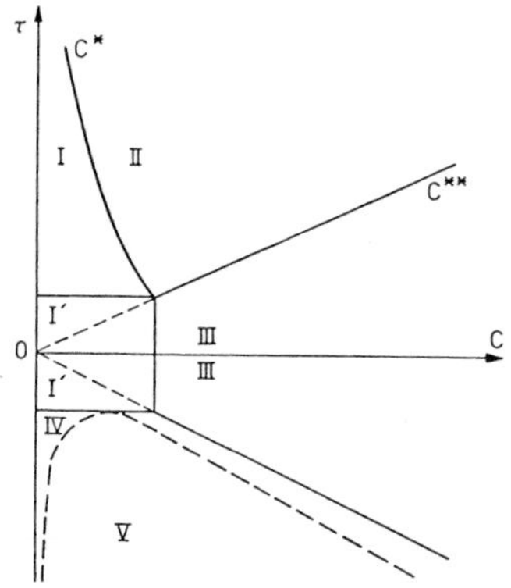

Fig.7.23

▲

Fig.7.22. Correlation diagram of characteristic oscillations in the IINS spectra of various hydrogen bronzes /7.68/

Fig.7.23. Temperature-concentration diagram of a polymer solution /7.70,71/. Ordinate: reduced temperature $\tau = (T-\Theta)/\Theta$, where Θ is the tricritical theta point. C = monomer concentration

In contrast to optical spectra IINS spectra can be described quantitatively by the scattering function, although in practice, as for QNS spectra, approximations are introduced to reduce the number of parameters. Thus the formulation of the scattering function for a harmonic (5.16) or anharmonic oscillator permits a satisfactory description of the IINS spectra of hydrogen in oxide bronzes /7.67,68/. However, neglecting the influence of lattice vibrations can lead to systematic errors in the calculated intensities, as shown in the example of H_xTaS_2 /6.4/. This point is discussed in detail in /7.52/.

7.6 Cross-Over Phenomena in Polymer Solutions

The thermodynamics of polymers in solution is more complex than the three models mentioned in Sect.6.3.1 would indicate. Figure 7.23 shows the recently derived (T,C) diagram /7.70,71/, which summarises the physical properties of polymer solutions as a function of temperature T and concentration C. Here the reduced temperature $\tau = (T-\Theta)/\Theta$ is chosen as the ordinate; C is the monomer concentration, Θ is the tricritical Θ point on the temperature scale.

A key to understanding the behavior of dissolved polymer molecules is the relationship between attraction and repulsion in the monomer-monomer interaction. While attraction is predominant at low temperatures (poor solvent behavior), so that the polymer molecule collapses (Region IV in Fig.7.23) or separation occurs (Region V), at high temperatures (good solvent behavior) repulsion is predominant because of the excluded volume (Regions I and II). The volume exclusion effect causes an expansion or "swelling" of the conformation distribution ("self-avoiding random walk" problem) and thus an increase in the radius of gyration R_G relative to a situation in which attraction and repulsion just cancel each other. The latter state can be achieved by an appropriate choice of solvent, temperature, and concentration, and is realised in the so-called Θ ranges (I' and III). Here the chain conformation is unperturbed and respects the Gaussian statistics ("random walk").

The lines between Regions I and II and I' and III, respectively, separate dilute ($C < C^*$) from semi-dilute solutions ($C > C^*$), where C^* is the critical concentration at which polymer coils of radius R_G just touch without penetrating each other.

7.6.1 Static Structure

Cross-over phenomena, which we wish to discuss here, occur in the vicinity of the "phase boundaries" of the temperature-concentration diagram (Fig.7.23). In each of the regions characteristic scaling laws for the variables temperature (T), concentration (C) and molecular weight (M_W) apply for the radius of gyration of the polymer solution (Sect.6.3). During the cross-over between two different regions, these scaling laws change continuously from one behavior to the other.

The mean square of the gyration radius $\langle R_G^2 \rangle$ is closely related to the second moment of the static monomer-monomer pair-correlation of the polymer molecule g(r), and thus to the static structure factor S(Q). It is well-known that $\langle R_G^2 \rangle$ is determined by measurement of S(Q) in small-angle scattering experiments /6.11/. The function S(Q) however contains more information than $\langle R_G^2 \rangle$; it is therefore interesting to study cross-over phenomena directly on S(Q).

Cross-over behavior is characterised by the fact that the transition from predominantly attractive to predominantly repulsive monomer-monomer interactions does not take place in the same way for all monomer pairs, but depends on their chemical distance n. If subsequent monomers in a chain molecule are numbered consecutively, the chemical distance between monomers (i,j) is defined as $n = (j-i)$, in contrast to the actual distance $r_n = |\underline{r}_j - \underline{r}_i|$. According to /7.71/, at a given reduced temperature τ in a dilute solution there is a characteristic distance

$$n_\tau = \tau^{-2} \quad , \tag{7.9}$$

whose significance is that all mean squares of the actual distances $\langle r_n^2 \rangle$ with $n < n_\tau$ are unperturbed and all $\langle r_n^2 \rangle$ with $n \geq n_\tau$ are swollen. Accordingly a temperature-dependent correlation length

$$\xi_\tau = \ell \cdot n_\tau^{1/2} = \ell / \tau \tag{7.10}$$

can be defined, in which ℓ is a segment length of the polymer chain. Now one property of the structure factor $S(\underline{Q})$, attributable to the Fourier transformations in (4.7, 43), is that essentially large-distance correlations determine its form in the range of small $\underline{Q}$ vectors, whereas small-distance correlations mainly have an effect in the large $\underline{Q}$ range. It may therefore be expected that in reciprocal space there is a cross-over length $Q^*_\tau \cong \xi_\tau^{-1}$ at which the behavior of S(Q) changes in a characteristic way. This change should reflect the transition from the excluded-volume interaction to Gaussian statistics on the scale of reciprocal lengths. As can be seen from the above equations, Q^*_τ is a temperature-dependent quantity. The effect is known as "temperature cross-over". A related phenomenon is predicted by the theory /7.70/ for solutions in good solvent in the semi-dilute region. Here the excluded-volume interaction is screened above a characteristic distance $n_c \geqq n_\tau$ because of the overlap with other polymer molecules, so that for $n > n_c$ Gaussian statistics prevail again. Thus chains consisting of n elements are swollen if $n < n_c$. The structure factor of a single chain then behaves as

$$S_1(Q) \propto Q^{-5/3} \quad , \tag{7.11}$$

whereas the structure factor for the (Gaussian) random coil configuration scales as

$$S_1(Q) \propto Q^{-2} \quad . \tag{7.12}$$

The exact formulation of $S_1(Q)$ describing this cross-over may be found in /7.71/. The transition between the two types of behavior is expected at a reciprocal cross-over length $Q^*_c \cong \xi_c^{-1}$, where the correlation length ξ_c is connected with n_c by the relation

$$\xi_c = \ell \cdot n_c^{3/5} \propto \ell \cdot C^{-3/4} \quad . \tag{7.13}$$

According to this, Q^*_c is a concentration-dependent quantity. This effect is known as "concentration cross-over".

These (static) cross-over phenomena predicted by the theory /7.70/ were investigated by small-angle neutron scattering in Regions I and II of the (T,C) diagram of polystyrene in CS_2 solution and their existence was confirmed /7.71/.

As an example Fig.7.24 shows results of the study of the concentration cross-over /7.71/. The measurements were carried out in "good" solution in the semi-dilute range. The reciprocal measured intensity is plotted in Fig.7.24 against $Q^{5/3}$ [cf. (7.11)]. The concentrations used are shown on the right; the measured points are compared with theory (continuous line). The agreement is excellent and the theoretical cross-over points are shown by vertical arrows. For $Q > Q^*_c$ the curve is a straight line. The values of Q^*_c increase with the concentration dependence $Q^*_c \propto C^{3/4}$ resulting from

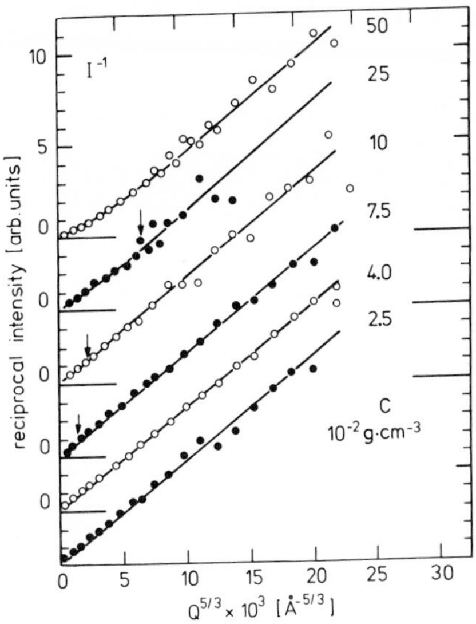

Fig.7.24

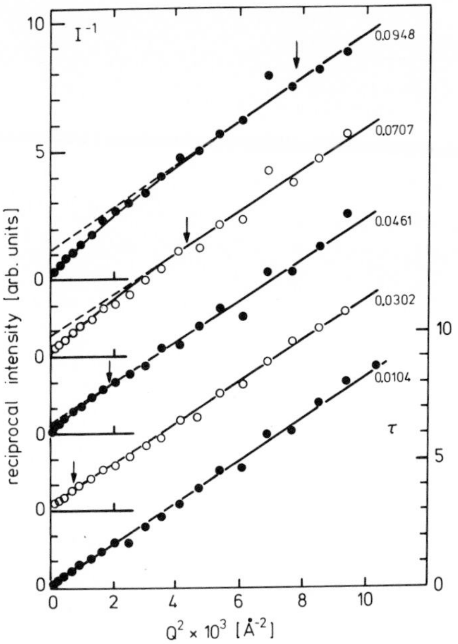

Fig.7.25

Fig.7.24. Static concentration cross-over for polystyrene in CS_2 solution. Open and solid circles represent the reciprocal measured intensity. Continuous lines correspond to the model calculations explained in the text for polymer molecules at the transition from Gaussian statistics to excluded-volume interaction. The calculated cross-over points are marked with arrows /7.71/

Fig.7.25. Static temperature cross-over for polystyrene in C_6D_{12} solution: reciprocal measured intensity as a function of Q^2 for reduced temperatures between τ = 0.0104 and 0.0948. Vertical arrows indicate theoretical cross-over points /7.71/

(7.13). As the concentration increases the reciprocal intensity slowly changes from the excluded-volume behavior (straight line) to that of Gaussian statistics (concave shape of the curve). A similarly good agreement with the theoretical predictions was found in the study of the temperature cross-over in polystyrene (PSH) in deuterated cyclohexane with $C = C^*/10$. The experiment was carried out in Region I of the (T,C) diagram (Fig.7.23). Figure 7.25 shows the reciprocal measured intensity as a function of Q^2 for reduced temperatures between $\tau = 0.0104$ and $\tau = 0.0948$ /7.71/. The continuous curves were obtained from the theory normalised to the experimental points. Vertical arrows indicate the theoretical cross-over points, which agree only qualitatively with the experimental values of Q^*_τ. These can be expressed by the following linear expression:

$$Q^*_\tau[\text{Å}^{-1}] = 1.606\,\tau + 0.0038 \quad . \tag{7.14}$$

Thus it can be said that the existence of the characteristic chemical distances n_τ and n_c has been proved using the measurement of the Q dependence of the structure factor $S_1(Q)$. When excluded-volume interaction is present, not only the parameters R_G and ℓ are required for the description of the chain conformation, but also the characteristic correlation lengths ξ_τ and ξ_c.

7.6.2 Dynamic Structure

We have already mentioned in Sect.6.4 an experiment concerning the dynamics of polymers, the aim of which was the qualitative identification of the internal modes of motion of the chain molecule in "liquid" and amorphous polymers. In view of the considerable number of available results, we limit ourselves here to the field of polymer solutions and in particular to that of segmental diffusion. Whereas local internal types of motion (e.g. of small side groups) are studied in the Q range of reciprocal bond lengths, the appropriate region for the study of conformation changes with time of polymer molecules is somewhat lower. It lies in fact between the reciprocal values of the segment length ℓ and the gyration radius R_G: $1/R_G \ll Q \ll 1/\ell$. This range is accessible to high-resolution neutron spin-echo spectroscopy, in which the real part of the intermediate scattering function $I(\underline{Q},t)$ (Sect.4.2) is determined directly by measuring the polarisation of the scattered neutrons as a function of the applied magnetic guide field /7.72,73/. The scattering functions (both incoherent and coherent) for segmental diffusion in dilute polymer solutions were calculated as early as 1967 /7.74/. Hydrodynamic equations were used as a basis for the interaction between the different segments and the excluded volume was not taken into account, so that the result really applies to the Θ range. As the coherent scattering function $S_{DG}(Q,\omega)$ appears more suitable for studies in the Q range of interest here because of the small-angle scattering involved, we shall limit the discussion to this function. The quasielastic line shape of $S_{DG}(Q,\omega)$ deviates considerably from that of a Lorentzian. For its half-width $\Delta\omega$ a universal behavior is predicted as a function of Q:

$$\Delta\omega \propto Q^3 \quad . \tag{7.15}$$

This scaling law is maintained qualitatively (i.e. apart from a constant factor), if excluded-volume effects are taken into account /7.75/, so that it should also be valid for "good" solutions. Neutron spin-echo experiments on segmental diffusion of polymers in good solvents have been carried out on polydimethyl siloxane (PDMS, $M_W = 3\cdot10^4$) and polymethyl methacrylate (PMMA, $M_W = 7\cdot10^6$) in deuterated benzene /7.76/. The theoretical function $S_{DG}(Q,\omega)$ gives a good description of the experimental data. The quasielastic linewidths $\Delta\omega$ were determined from a fit of the corresponding intermediate scattering function $I_{DG}(Q,t)$. In Fig.7.26 these are plotted

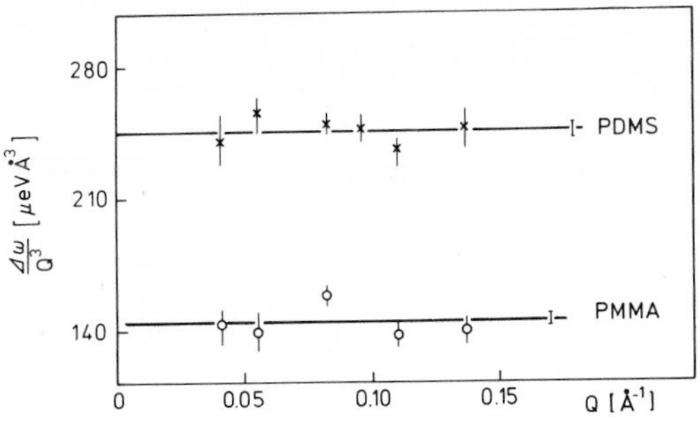

Fig.7.26. Quasielastic linewidth $\Delta\omega$ of polydimethyl siloxane (PDMS) and polymethyl methacrylate (PMMA) in deuterated benzene measured with the neutron spin-echo technique: confirmation of the Q^3 scaling behavior /7.76/

against Q after division by Q^3 /7.76/. The Q^3 scaling behavior predicted by the theory is thus impressively demonstrated. In the same investigation the dynamic concentration cross-over theoretically predicted in /7.75/ was observed at higher polymer concentrations. As in the static case the theory defines a correlation length

$$\xi_c \propto c^{-3/4} \quad . \tag{7.16}$$

The single-chain behavior applies for $Q\xi_c \geqq 1$ with $\Delta\omega \propto Q^3$, whereas for $Q\xi_c < 1$ collective diffusion predominates with

$$\Delta\omega \cong D_c Q^2 \quad . \tag{7.17}$$

Here D_c is the cooperative diffusion constant of the gel. In Fig.7.27 we show the results of measurements of the concentration cross-over of PDMS solutions in deuterated benzene /7.76/, where the cross-over from Q^3 to Q^2 behavior was indeed observed. With increasing concentration this takes place at higher Q values. Above each cross-over all the linewidth values lie on the same straight line. At small Q the diffusion constant D_c is obtained directly from the height of the plateau. These results represent the first direct observation of the spatial transition from single-chain to multichain behavior.

In /7.75,77/, the dynamics of polymer solutions is discussed from the viewpoint of the scaling laws. These laws predict a simple universal behavior of polymer para-

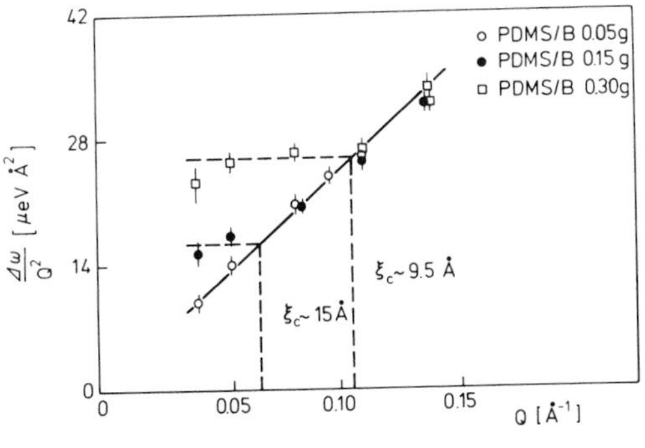

Fig.7.27. Study of the dynamic concentration cross-over (Fig.7.23) of PDMS in deuterated benzene by the neutron spin-echo technique. The cross-over from Q^3 scaling (corresponding to single-chain behavior) to Q^2 scaling (multi-chain behavior) occurs with increasing concentration at higher Q values (see text) /7.76/

meters in certain ranges of Q, T and C. However, they do not give any precise values for coefficients and give no information on the cross-over region where these simple laws do not apply. This type of information requires the solution of dynamic equations. Starting from a generalised (Fokker-Planck) diffusion equation, it is possible to calculate the derivative of the logarithm of the intermediate scattering function in the $t\to 0$ limit

$$\Omega(Q) = \lim_{t\to 0} \frac{\partial}{\partial t} \ln I(Q,t) \qquad (7.18)$$

from static correlation functions alone /7.78/. As $\Omega(Q)$ is proportional to the half-width $\Delta\omega$ of $S(Q,\omega)$, the behavior of $\Delta\omega$ can thus be studied numerically, for instance in the temperature cross-over region between I' and I. Figure 7.28 shows the result of such a calculation /7.79/. Here the reduced initial derivative $\Omega/(TQ^3/\eta)$ is plotted for two different chain lengths N against the reduced temperature for different values of $Q\ell$; η is the viscosity of the solvent. According to this the theory predicts the following behavior for the dynamic temperature cross-over:

1. The cross-over is observed as a step in Ω and thus in the quasielastic line-width.
2. The cross-over from Gaussian to excluded-volume dynamics is the more marked, and the step is the higher, the smaller $Q\ell$ and the longer the chains (i.e. the greater the value of N).
3. As $Q\ell$ increases, the cross-over point is shifted towards higher temperatures.

These predictions have been qualitatively confirmed by experiment. The system polystyrene ($M_W \cong 60000$)/deuterated cyclohexane, polymer concentration 5%, already stud-

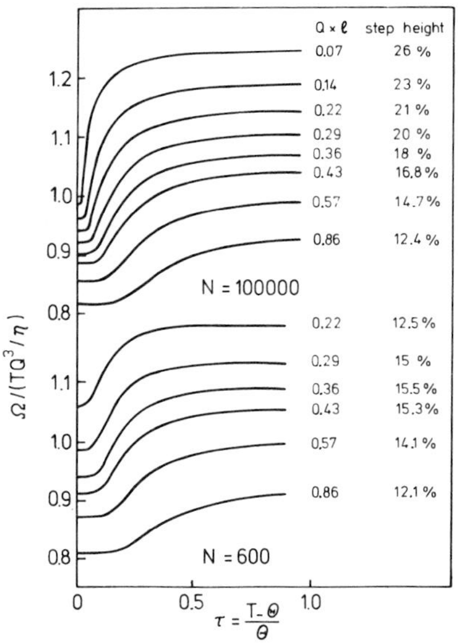

Fig.7.28. Dynamic temperature cross-over. Calculation of $\Omega/(TQ^3/\eta)$ for two chain lengths N as a function of the reduced temperature (Fig. 7.23). Here η is the viscosity of the solvent; $\Omega(Q)$ is proportional to the half-width $\Delta\Omega$ of $S(Q,\omega)$. The cross-over is observed as a step in Ω and hence in $\Delta\omega$ /7.79/

ied in static measurements /7.71/, was investigated in a spin-echo experiment /7.79/. Figure 7.29 shows the experimental results similarly to Fig.7.28; the qualitative agreement of the two illustrations is evident. This is clear experimental proof of the existence of a dynamic temperature cross-over effect in dilute solutions. Some questions are still open from the quantitative viewpoint. Thus the observed step is much higher and much more marked than that theoretically predicted, and the observed Q dependences of step height and step shape are much more pronounced. In addition, static and dynamic cross-over points do not agree quantitatively.

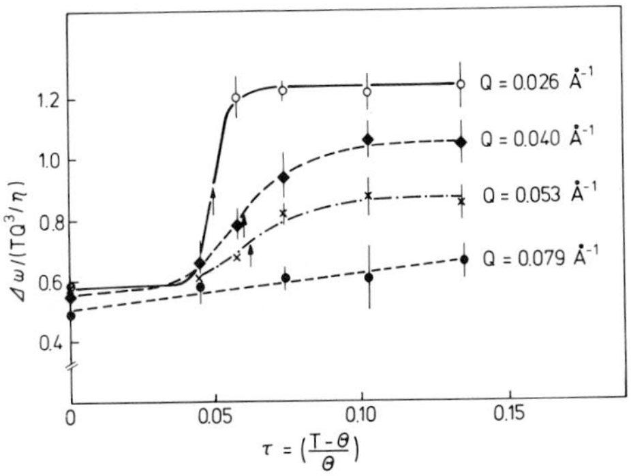

Fig.7.29. Experimental verification of the dynamic temperature cross-over for a solution of polystyrene ($M_W \cong 60000$) in deuterated cyclohexane, polymer concentration 5% (Fig.7.28) /7.79/

8. Conclusion

In the present volume we have attempted to show that the neutron scattering technique has developed in recent years from being a special domain of the physicist into a versatile tool able to make important contributions to the solution of topical problems in many areas of research, particularly in chemistry where it is in the course of becoming a routine technique. One of the preconditions for this development was the creation of the Institut Laue-Langevin (ILL) at Grenoble with the specific aim of developing the most modern methods of neutron scattering and making these available to the international community. Thus the ILL has become a pioneer of a new generation of intense neutron sources which will come into operation in the 1980s in the United Kingdom and the United States. In West Germany, too, there are plans to extend neutron scattering research facilities.

Appendix A: Scattering Function and Thermal Average

To discuss the transition from (4.12) to (4.13) in more detail, we consider the double sum

$$Y = \sum_{\psi_0} p(\psi_0) \sum_{\psi_1} \langle\psi_0|e^{-i\underline{Q}\underline{r}_j}|\psi_1\rangle\langle\psi_1|e^{i\underline{Q}\underline{r}_i}|\psi_0\rangle \cdot \exp\left[\frac{i}{\hbar}(E_{\psi_1}-E_{\psi_0})t\right] \quad . \tag{A.1}$$

The closure relation – applied to (A.1) – indicates that for any pair of operators A and B the summation over all final states ψ_1 gives

$$\sum_{\psi_1} (\int\psi_0^* A\psi_1 d\underline{r})(\int\psi_1^* B\psi_0 d\underline{r}) = \int\psi_0^* AB\psi_0 d\underline{r} \quad , \tag{A.2}$$

where the asterix means complex conjugated. In Dirac notation the same equation reads:

$$\sum_{\psi_1} \langle\psi_0|A|\psi_1\rangle\langle\psi_1|B|\psi_0\rangle = \langle\psi_0|AB|\psi_0\rangle \quad . \tag{A.3}$$

Equation (A.1) thus simplifies to

$$Y = \sum_{\psi_0} p(\psi_0)\langle\psi_0|e^{-i\underline{Q}\underline{r}_j}\cdot e^{i\underline{Q}\underline{r}_i}|\psi_0\rangle \exp\left[\frac{i}{\hbar}(E_{\psi_1}-E_{\psi_0})t\right] \quad . \tag{A.4}$$

Now the last factor of the product can stand anywhere; we may for example write:

$$Y = \sum_{\psi_0} p(\psi_0)\langle\psi_0|e^{-i\underline{Q}\underline{r}_j}\cdot\exp(iE_{\psi_1}t/\hbar)\, e^{i\underline{Q}\underline{r}_i}\exp(-iE_{\psi_0}t/\hbar|\psi_0\rangle \tag{A.5}$$

To simplify (A.5) further we use the definition of the time-dependent Heisenberg operator $\underline{r}_j(t)$:

$$\underline{r}_j(t) = \exp(iHt/\hbar)\underline{r}_j(0)\exp(-iHt/\hbar) \quad , \tag{A.6}$$

from which (by expanding the exponential function) it follows that

$$\exp[i\underline{Q}\underline{r}_j(t)] = \exp(iHt/\hbar)\exp[i\underline{Q}\underline{r}_j(0)]\exp(-iHt/\hbar) \quad . \tag{A.7}$$

The wave functions ψ_0 and ψ_1 are eigenfunctions with eigenvalues E_{ψ_0} and E_{ψ_1} of the Hamilton operator H of the scattering system, i.e. according to the Schrödinger equation we have

$$H|\psi_0\rangle = E_{\psi_0}|\psi_0\rangle \quad \text{and} \quad H|\psi_1\rangle = E_{\psi_1}|\psi_1\rangle \quad . \tag{A.8}$$

If H is applied n times to $|\psi_0\rangle$ then

$$H^n|\psi_0\rangle = E_\psi^n|\psi_0\rangle \quad , \tag{A.9}$$

so that clearly (as can be seen by expanding the exponential function)

$$\exp(-iHt/\hbar)|\psi_0\rangle = \exp(-iE_{\psi_0}t/\hbar)|\psi_0\rangle \quad . \tag{A.10}$$

Inserting (A.10) into (A.5), with the use of (A.6), yields

$$Y = \sum_{\psi_0} p(\psi_0)\langle\psi_0|e^{-i\underline{Q}\underline{r}_j(0)}\, e^{i\underline{Q}\underline{r}_i(t)}|\psi_0\rangle \quad . \tag{A.11}$$

[Note: by definition $\underline{r}_j(0) = \underline{r}_j$.]

Comparison of (A.11) with the definition of the thermal average of an operator A at temperature T, i.e. with

$$\langle A\rangle = \sum_{\psi_0} p(\psi_0)\langle\psi_0|A|\psi_0\rangle \quad , \tag{A.12}$$

finally gives the desired result:

$$Y = \langle e^{-i\underline{Q}\underline{r}_j(0)}\, e^{i\underline{Q}\underline{r}_i(t)}\rangle \quad . \tag{A.13}$$

Appendix B: Application of the Convolution Theorem

The transition from (4.22) to (4.23) requires recasting of the integral

$$Y(\underline{r}) = \int e^{-i\underline{Q}\underline{r}}\{\exp[-i\underline{Q}\underline{r}_j(0)]\exp[i\underline{Q}\underline{r}_i(t)]\}d\underline{Q} \quad . \tag{B.1}$$

This is the Fourier transformation of a product. In abbreviated form the same equation may be written

$$Y(\underline{r}) = \int e^{-i\underline{Q}\underline{r}}X(\underline{Q})d\underline{Q} = \int e^{-i\underline{Q}\underline{r}}[X_1(\underline{Q})X_2(\underline{Q})]d\underline{Q} \quad . \tag{B.2}$$

If we define the Fourier transforms

$$Y_1(\underline{r}) = \int e^{-i\underline{Q}\underline{r}}X_1(\underline{Q})d\underline{Q} \tag{B.3}$$

and

$$Y_2(\underline{r}) = \int e^{-i\underline{Q}\underline{r}}X_2(\underline{Q})d\underline{Q} \tag{B.4}$$

of the factors X_1 and X_2, the convolution theorem for a product states that the Fourier transform of the latter is equal to the convolution integral of the individual transforms Y_1 and Y_2:

$$Y(\underline{r}) = \int d\underline{r}'Y_1(\underline{r}-\underline{r}')Y_2(\underline{r}') \quad . \tag{B.5}$$

Explicitly we have:

$$Y_1(\underline{r}-\underline{r}') = \int e^{-i\underline{Q}(\underline{r}-\underline{r}')} \exp[-i\underline{Q}\underline{r}_j(0)]d\underline{Q} \tag{B.6}$$
$$= (2\pi)^3\, \delta[\underline{r}+\underline{r}_j(0)-\underline{r}']$$

and

$$Y_2(\underline{r}') = \int e^{-i\underline{Q}\underline{r}'} \exp[i\underline{Q}\underline{r}_i(t)]d\underline{Q} = (2\pi)^3\, \delta[\underline{r}'-\underline{r}_i(t)] \quad , \tag{B.7}$$

from which (4.23) follows immediately by insertion into (B.5) and (4.22).

Acknowledgement. The authors are obliged to Prof. Dr. T. Springer for encouraging them to write this article and to Prof. Dr. E. Wicke for several valuable suggestions. We thank Prof. Dr. M. Zeidler for a critical reading of the manuscript and a number of colleagues for useful discussions and remarks concerning parts of the text, especially: Prof. Dr. H. Dachs, Dr. H. Hässlin, Dr. H. Lauter, Dr. M.S. Lehmann, Dr. R. May, Dr. R. Oberthür, Dr. W. Press, Dr. D. Richter, Dr. W. Weppner, and Dr. G. Zaccai.

References

1.0 B. Dorner, H. Peisl: Nucl. Instr. Meth. *208*, 587 (1983), and refs. therein
1.1 H. Boutin, S. Yip: *Molecular Spectrosopy with Neutrons* (The M.I.T. Press, Cambridge, Mass. and London, England 1968)
1.2 B.T.M. Willis (ed.): *Chemical Applications of Thermal Neutron Scattering* (Oxford University Press, London 1973)
1.3 G.E. Bacon: *Neutron Scattering in Chemistry* (Butterworth Ltd., London 1977)
1.4 S.W. Lovesey, T. Springer (ed.): "Dynamics of Solids and Liquids by Neutron Scattering", *Topics in Current Physics*, Vol. 3 (Springer, Berlin, Heidelberg, New York 1977)
1.5 H. Dachs (ed.): "Neutron Diffraction", *Topics in Current Physics*, Vol. 6 (Springer, Berlin, Heidelberg, New York 1978)
1.6 G. Kostorz (ed.): "Neutron Scattering in Materials Science", in: *A Treatise on Materials Science and Technology*, Vol. 15, Series Editor H. Herman (Academic Press, New York 1979)
1.7 H. Fuess: "Application of Neutron Diffraction to Chemistry", in *Modern Physics in Chemistry*, ed. by E. Fluck, V.I. Goldanskii (Academic Press, London 1978)
1.8 R.M. Moon (ed.): Proc. Conf. Neutron Scattering, Gatlinburg, June 1976, available from Natl. Techn. Info. Service, U.S. Dept. Commerce, Springfield, Va. 22161 (1976)
1.9 *Neutron Inelastic Scattering 1977*, Vol. I and II (IAEA Vienna 1978)
2.1 P.A. Egelstaff (ed.): *Thermal Neutron Scattering* (Academic Press, London, New York 1965)
2.2 G.E. Bacon: *Neutron Diffraction*, 3rd Edition (Clarendon Press, Oxford 1975)
2.3 B. Maier (ed.): "Neutron Beam Facilities at the HFR available for Users", Internal Report (Institut Laue-Langevin, Grenoble 1977)
3.1 L. Koester: "Neutron Scattering Lenghts and Fundamental Neutron Interactions", in: *Springer Tracts in Modern Physics*, Vol. 80 (Springer, Berlin, Heidelberg, New York 1977) pp. 1-56
4.1 T. Springer: "Quasielastic Neutron Scattering for the Investigation of Diffusive Motions in Solids and Liquids", in: *Springer Tracts in Modern Physics*, Vol. 64 (Springer, Berlin, Heidelberg, New York 1972)

4.2 I.I. Gurevich, L.V. Tarasov: *Low-Energy Neutron Physics* (North-Holland, Amsterdam 1968)
4.3 W. Marshall, S.W. Lovesey: *Theory of Thermal Neutron Scattering* (Clarendon, Oxford 1971)
4.4 G.L. Squires: *Introduction to the Theory of Thermal Neutron Scattering* (Cambridge University Press, Cambridge 1978)
4.5 A. Hüller: Phys. Rev. B *16*, 1844 (1977)
4.6 A. Hüller, W. Press: in ref. /1.9/ Vol. 1, p. 231
4.7 L. Van Hove: Phys. Rev. *95*, 249 (1954)
4.8 L. Blum, A.N. Narten: Adv. Chem. Phys. *34*, 203 (1976)
4.9 R.E. Lechner: "Neutron Scattering Studies of Diffusion in Solids", Chapter 8, in: *Mass Transport in Solids*, NATO-ASI Series, ed. by F. Bénière, R. Catlow (Plenum Press, New York 1983) pp. 169-226
5.1 C.H. Vineyard: Phys. Rev. *110*, 999 (1958)
5.2 C.T. Chudley, R.J. Elliott: Proc. Phys. Soc. (London) *77*, 353 (1961)
5.3 J.M. Rowe, K. Sköld, H.E. Flotow, J.J. Rush: J. Phys. Chem. Sol. *32*, 41 (1971)
5.4 J.N. Sherwood (ed.): *The Plastic Crystalline State* (John Wiley, New York 1979)
5.5 V.F. Sears: Can. J. Phys. *45*, 237 (1967)
5.6 R. Stockmeyer, H. Stiller: Phys. Stat. Sol. *27*, 269 (1968)
5.7 K. Sköld: J. Chem. Phys. *49*, 2443 (1968)
5.8 R.E. Lechner, J.M. Rowe, K. Sköld, J.J. Rush: Chem. Phys. Lett. *4*, 444 (1969)
5.9 A.J. Leadbetter, R.E. Lechner: in ref. /5.4/, pp. 285-320
5.10 B. Dorner, R. Comès: in ref. /1.4/, p. 127
5.11 B. Dorner: "Coherent Inelastic Neutron Scattering in Lattice Dynamics", in: *Springer Tracts in Modern Physics*, Vol. 93 (Springer, Berlin, Heidelberg, New York 1982)
6.1 I. Olovsen and P.G. Jönssen: in: "The Hydrogen Bond", Vol. II, ed. by P. Schuster, G. Zundel, C. Sandorfy (North-Holland, Amsterdam 1976) p. 393
6.2 R. Bau, R.G. Teller, S.W. Kirtley, T.F. Koetzle: Acc. Chem. Res. *12*, 176 (1979)
6.3 G. Zaccai: in ref. /1.5/, p. 243
6.4 C. Riekel, H.G. Reznik, R. Schöllhorn, C.J. Wright: J. Chem. Phys. *70*, (11), 5203 (1979)
6.5 F.E. Massoth: in: "Advances in Catalysis and Related Subjects", Vol. 27, ed. by D.D. Eley, H. Pines, P.B. Weisz (Academic Press, New York 1978)
6.6 A. Mitschler, B. Rees, M.S. Lehmann: J. Am. Chem. Soc. *100*, 3390 (1978)
6.7 P. Coppens: in ref. /1.5/, p. 71
6.8 P.J. Brown, A. Capiomont, B. Gillon, J. Schweizer: J. of Magnetism and Magnetic Materials *14*, 289 (1979)
6.9 A.W. Salotto, L. Burnelle: J. Chem. Phys. *53*, 333 (1970)
6.10 Y. Ellinger: "Etude théorique des relations entre structure et propriétés physico-chimiques dans les séries nitroxide et cetyle", PhD Thesis, University of Grenoble (1973)
6.11 A. Guinier, G. Fourney: *Small Angle Scattering of X-Rays* (W.H. Freeman, Brighton 1955)
6.12 P.J. Flory: *Statistical Mechanics of Chain Molecules* (Interscience, New York 1969)
6.13 P.G. De Gennes: *Scaling Concepts in Polymer Physics* (Cornell University Press, Ithaca 1979)
6.14 H. Benoit, D. Decker, J.S. Higgins, C. Picot, J.P. Cotton, B. Farnoux, G. Jannink, R. Ober: Nature Phys. Sc. *245*, 13 (1973)
6.15 B. Jacrot: Rep. Prog. Phys. *39*, 911 (1976)
6.16 H.B. Stuhrmann: Chemie Unserer Zeit *11*, 1 (1976)
6.17 G. Zaccai, P. Morin, B. Jacrot, D. Moras, J.C. Thierry, R. Giegé: J. Mol. Biol. *129*, 483 (1979)
6.18 H. Hervet, A.J. Dianoux, R.E. Lechner, F. Volino: J. Phys. (Paris) *37*, 587 (1976)
6.19 G. Allen, J.S. Higgins, C.J. Wright: J. Chem. Soc. Faraday Trans. II, *70*, 348 (1974)

6.20 G. Allen, J.S. Higgins: Macromolecules *10*, 1006 (1977)
6.21 H. Hervet, F. Volino, A.J. Dianoux, R.E. Lechner: J. de Phys. Lettres *35*, L-151 (1974)
6.22 H. Hervet, F. Volino, A.J. Dianoux, R.E. Lechner: Phys. Rev. Lett. *34*, 451 (1975)
6.23 A.J. Dianoux, F. Volino: in ref. /1.9/, Vol. 1, p. 533
6.24 R.E. Lechner, A. Heidemann: Commun. Phys. *1*, 213 (1976)
6.25 J. Töpler, D.R. Richter, T. Springer: J. Chem. Phys. *69*, 3170 (1978)
6.26 R.W. Gerling: "Rotational Motion of Molecules in Condensed Phases", PhD. Thesis, University of Erlangen (1981); Jül-1726, July 1981, ISSN0366-0885 (Kernforschungsanlage Jülich 1981)
6.27 M. Bée, J.P. Amoureux, R.E. Lechner: Mol. Phys. *40*, 617 (1980)
6.28 R.E. Lechner: in ref. /1.8/, Vol. I, p. 310
6.29 H. Kapulla, W. Gläser: in: "Inelastic Scattering of Neutrons in Solids and Liquids" (IAEA, Vienna 1973) p. 841
6.30 W. Press, A. Kollmar: Solid State Commun. *17*, 405 (1975)
6.31 M. Prager, W. Press, B. Alefeld, A. Hüller: J. Chem. Phys. *67*, 5126 (1977)
6.32 W. Press: "Single Particle Rotations in Molecular Crystals", in: *Springer Tracts in Modern Physics*, Vol. 92 (Springer, Berlin, Heidelberg, New York 1981)
6.33 K. Sköld: in: *Topics in Applied Physics*, Vol. 28, "Hydrogen in Metals I. Basic Properties", ed. by G. Alefeld, J. Völkl (Springer, Berlin, Heidelberg, New York 1978) p. 267
6.34 K. Sköld, G. Nelin: J. Phys. Chem. Solids *28*, 2369 (1967)
6.35 J.M. Rowe, J.J. Rush, L.A. de Graaf, G.A. Ferguson: Phys. Rev. Lett. *29*, 1250 (1972)
6.36 D. Richter, J. Töpler, T. Springer: J. Phys. F *6*, L93 (1976)
6.37 D. Richter, T. Springer: Phys. Rev. B *18*, 126 (1978)
6.38 G. Göltz, A. Heidemann, H. Mehrer, A. Seeger, D. Wolf: Phil. Mag. A *41*, 723 (1980)
6.39 L.A. de Graaf, J.J. Rush, H.E. Flotow, J.M. Rowe: J. Chem. Phys. *56*, 4574 (1972)
6.40 D. Richter, B. Alefeld, A. Heidemann, N. Wakabayashi: J. Phys. F *7*, 569 (1977)
6.41 H. Rauch, A. Zeilinger: "Atomic Energy Review", Vol. 15, No. 2, p. 249, IAEA Vienna (1977)
7.1 R. Allemand, J. Bourdel, P. Convert, J.P. Cotton, B. Farnoux, K. Ibel, J. Jacobé: Nucl. Instrum. Methods *126*, 29 (1975)
7.2 F.C. Tompkins: in: *Reactivity in Solids*, Vol. 4 of "Treatise on Solid State Chemistry", ed. by N.B. Hannay (Plenum Press, New York 1976)
7.3 A.F. Wright, J. Talbot, B.E.F. Fender: Nature *277*, 366 (1979)
7.4 F.R. Gamble, T.H. Geballe: *Treatise on Solid State Chemistry*, Vol. 3, ed. by N.B. Hannay (Plenum Press, New York 1975)
7.5 R. Schöllhorn: Angew. Chemie *92*, 1015 (1980)
7.6 C. Riekel, H.G. Reznik, R. Schöllhorn: J. Solid State Chem. *29*, 181 (1979)
7.7 W. Metz, D. Hohlwein: Carbon *13*, 87 (1973)
7.8 C. Riekel: in: "Prog. Solid State Chem.", ed. by G.M. Rosenblatt, W.L. Worrell, *13*, 89 (1980)
7.9 C. Riekel, H.G. Reznik, R. Schöllhorn: unpublished
7.10 J.W. White, A.D. Taylor, J.C. Lassegues: in: "Protons and Ions Involved in Fast Dynamic Phenomena", ed. by. P. Laszlo (Elsevier Publ. Co., Amsterdam 1978) p. 123
7.11 R.E. Lechner, J.P. Amoureux, M. Bée, R. Fouret: Commun. Phys. *2*, 207 (1977)
7.12 J.P. Amoureux, M. Bée, R. Fouret, R.E. Lechner: in ref. /1.9/, Vol. I, p. 397
7.13 M. Bée, J.P. Amoureux, R.E. Lechner: Mol. Phys. *39*, 945 (1980)
7.14 M.B. Salamon (ed.): "Physics of Superionic Conductors", *Topics in Current Physics*, Vol. 15 (Springer, Berlin, Heidelberg, New York 1979)
7.15 P. Hagenmüller, W. van Gool (ed.): *Solid Electrolytes General Principles, Characterization, Materials, Applications* (Academic Press, New York (1977)
7.16 M. Ait-Salem, T. Springer, A. Heidemann, B. Alefeld: Phil. Mag. A *39*, 797 (1979)
7.17 W. Bührer, W. Hälg: Helv. Phys. Acta *47*, 27 (1974)

7.18 A.F. Wright, B.E.F. Fender: J. Phys. C *10*, 2261 (1977)
7.19 R.J. Cava, F. Reidinger, B.J. Wuensch: Solid State Commun. *24*, 411 (1977)
7.20 G. Eckold, K. Funke, J. Kalus, R.E. Lechner: J. Phys. Chem. Solids *37*, 1097 (1976)
7.21 R.E. Lechner, G. Eckold, K. Funke: in: "Microscopic Structure and Dynamics of Liquids", ed. by. J. Dupuy, A.J. Dianoux (Plenum Press, New York 1978) p. 459
7.22 P. Vashishta, A. Rahman: Phys. Rev. Lett. *40*, 1337 (1978)
7.23 P. Aldebert, A.J. Dianoux, J.P. Traverse: J. Phys. (Paris) *40*, 1005 (1979)
7.24 A.M. Thomy, X. Duval, J. Regnier: Surf. Sci. Reports *1*, 1 (1981)
7.25 H. Taub, K. Carneiro, J.K. Kjems, L. Passell, J.P. McTague: Phys. Rev. B *16*, 4551 (1977)
7.26 K. Carneiro: J. Phys. (Paris), Colloque No. 4, Fasc. C4-1 (1977)
7.27 B.E. Warren: Phys. Rev. *59*, 693 (1941)
7.28 J.K. Kjems, L. Passell, H. Taub, J.G. Dash, A.D. Novaco: Phys. Rev. B *13*, 1446 (1976)
7.29 M. Nielsen, J.P. McTague, W. Ellenson: J. Phys. (Paris), Colloque No. 4, Fasc. C4-10 (1977)
7.30 P. Vora, S.K. Sinha, R.K. Crawford: Phys. Rev. Lett. *43*, 704 (1979)
7.31 A.D. Novaco, J.P. McTague: Phys. Rev. Lett. *38*, 1286 (1977)
7.32 C.G. Shaw, S.C. Fain, Jr., M.D. Chinn: Phys. Rev. Lett. *41*, 955 (1978)
7.33 R.J. Birgeneau, E.M. Hammonds, P. Heiney, P.W. Stephens, P.M. Horn: in ref. /7.37/
7.34 J.M. Kosterlitz, D.J. Thouless: J. Phys. C *6*, 1181 (1973)
7.35 M. Bienfait: Surf. Sci. *89*, 13 (1979)
7.36 J. Villain: in: "Order in Strongly Fluctuating Condensed Matter Systems", ed. by T. Riste (Plenum, New York 1980) p. 221
7.37 S.K. Sinha (ed.): *Ordering in Two Dimensions* (North Holland, Amsterdam 1980)
7.38 Y. Larher, P. Thorel, B. Gilquin, B. Croset, Cl. Marti: Surf. Sci. *85*, 94 (1979)
7.39 Cl. Marti, B. Croset, P. Thorel, J.P. Coulomb: Surf. Sci. *65*, 532 (1977)
7.40 H. Taub, H.R. Danner, Y.P. Sharma, H.L. Mc.Murray, R.M. Brugger: Phys. Rev. Lett. *39*, 215 (1977)
7.41 J.P. Beaufils, Y. Barbaux: J. de Chim. Phys. *78*, 347 (1981)
7.42 J.P. Coulomb, J.P. Biberain, J. Suzanne, A. Thomy, G.J. Trott, H. Taub, H.R. Danner, F.Y. Hansen: Phys. Rev. Lett. *43*, 1878 (1979)
7.43 H.M. Rietveld: J. Appl. Crystallogr. *2*, 65 (1969)
7.44 G.J. Trott, H. Taub, F.Y. Hansen, H.R. Danner: Chem. Phys. Lett. *78*, 504 (1981)
7.45 M. Monkenbusch, R. Stockmeyer: Ber. Bunsenges. Phys. Chem. *84*, 808 (1980)
7.46 J.P. Coulomb, M. Bienfait, P. Thorel: J. Phys. (Paris), Colloque N. 4, Fasc. C4-31 (1977)
7.47 J.P. Coulomb, M. Bienfait, P. Thorel: Phys. Rev. Lett. *42*, 733 (1979)
7.48 M.W. Newbery, T. Rayment, M.V. Smalley, R.K. Thomas, J.W. White: Chem. Phys. Lett. *59*, 461 (1978)
7.49 D.M. Kolb: in: "Advances in Electrochemistry and Electrochemical Engineering", Vol. XI, ed. by H. Gerischer, C.W. Tobias (J. Wiley, New York 1978) p. 125
7.50 G. Bomchil, C. Riekel: J. Electroanal. Chem. Interfacial Electrochem. *101*, 133 (1979)
7.51 A.J. Renouprez, P. Fouilloux, J.P. Candy, J. Tomkinson: Surf. Sci. *83*, 285 (1979)
7.52 C.J. Wright, C.M. Sayers: Rep. Prog. Phys. *46*, 773 (1983)
7.53 H. Tributsch: Solar Energy Materials *1*, 705 (1979)
7.54 M.S. Whittingham: J. Solid State Chem. *29*, 303 (1979)
7.55 A.R. Bêal, W.Y. Liang: J. Phys. C *6*, L482 (1973)
7.56 C. Riekel, D. Hohlwein, R. Schöllhorn: J.C.S. Chem. Commun. 863 (1976)
7.57 C.J. Wright, B. Tofield: Solid State Commun. *22*, 715 (1977)
7.58 R. Schöllhorn, H.D. Zagefka, T. Butz, A. Lerf: Mater. Res. Bull. *14*, 369 (1979)
7.59 C. Riekel, C.O. Fischer: J. Solid State Chem. *29*, 181 (1979)
7.60 R. Schöllhorn, H.D. Zagefka: Angew. Chem. *3*, 193 (1977)
7.61 R. Schöllhorn, E. Sick, A. Lerf: Mater. Res. Bull. *10*, 1005 (1975)

7.62 C. Riekel, R. Schöllhorn, J. Tomkinson: Z. Naturforsch. *35a*, 590 (1980)
7.63 C. Riekel, A. Heidemann, B.E.F. Fender, G.C. Stirling: J. Chem. Phys. *71*, 530 (1979)
7.64 B.G. Silbernagel, M.B. Dines, F.R. Gamble, L.A. Gebhard, M.S. Whittingham: J. Chem. Phys. *65*, 1906 (1976)
7.65 C. Riekel: Solid State Commun. *28*, 385 (1979)
7.66 J.M. Adams, C. Breen, C. Riekel: J. Colloid Interface Sci. *68*, 214 (1979)
7.67 C.J. Wright, C. Riekel, R. Schöllhorn, B.C. Tofield: J. Solid State Chem. *24*, 219 (1978)
7.68 P.G. Dickens, J. Birtill, C.J. Wright: J. Solid State Chem. *28*, 185 (1979)
7.69 A. Cirillo, J.J. Fripiat: J. Phys. (Paris) *39*, 247 (1978)
7.70 M. Daoud, G. Jannink: J. Phys. (Paris) *37*, 973 (1976)
7.71 B. Farnoux, F. Boué, J.P. Cotton, M. Daoud, G. Jannink, M. Nierlich, P.G. De Gennes: J. Phys. (Paris) *39*, 77 (1978)
7.72 F. Mezei: Z. Phys. *255*, 146 (1972)
7.73 F. Mezei (ed.): "Neutron Spin Echo", *Lecture Notes in Physics*, Vol. 128 (Springer, Berlin, Heidelberg, New York 1980)
7.74 E. Dubois-Violette, P.G. De Gennes: Physics *3*, 181 (1967)
7.75 P.G. De Gennes: Macromolecules *9*, 594 (1976)
7.76 D. Richter, J.B. Hayter, F. Mezei, B. Ewen: Phys. Rev. Lett. *41*, 1484 (1978)
7.77 M. Daoud, G. Jannink: J. Phys. (Paris) *39*, 331 (1978)
7.78 M. Benmouna, A.Z. Akcasu: Macromolecules *11*, 1187 (1978)
7.79 D. Richter, B. Ewen, J.B. Hayter: Phys. Rev. Lett. *45*, 2121 (1980)

Abbreviations

A-A	adsorbate-adsorbate
A-S	adsorbate-substrate
EISF	elastic incoherent structure factor
FWHM	full-width-at-half-maximum
IINS	incoherent inelastic neutron spectroscopy
IR	infra-red
LEED	low-energy electron diffraction
mRNA	messenger ribonucleic acid
NMR	nuclear magnetic resonance
ODIC	orientationally disordered crystals
PAA	para-azoxy anisole
PDMS	polydimethyl siloxane
PMMA	polymethyl methacrylate
PSD	position-sensitive detector
QNS	quasielastic neutron scattering
RS	aminoacyl-tRNA-synthetase
SCCCMO	self-consistent charge calculation of molecular orbitals
TBBA	terephthal-bis-butyl aniline
TDS	thermal-diffuse scattering
TP	trigonally prismatic
tRNA	transfer ribonucleic acid
2D	two-dimensional
ValRS	valyl-tRNA synthetase

Transport Mechanisms of Light Interstitials in Metals

By D. Richter

1. Introduction

The transport phenomena of light interstitials in metals as hydrogen, deuterium, tritium and muon constitute a fascinating area of research which has already attracted experimental and theoretical efforts for a considerable amount of time. The hydrogen isotopes cover nearly a factor of 30 in mass and the whole range between classical hopping and electron-like band motion. Three main aspects appear to be prominent: (i) the question as to the *nature of the elementary diffusive step* is of prime interest for diffusion theories. Here, the muon as a particle at the borderline between bandlike propagation and quantum hopping seems to be the ideal probe to investigate the different features of quantum transport theory; (ii) *the space and time evolution of the diffusive process* can be studied in most detail by quasielastic neutron scattering, which offers a probe sensitive to the motion of an atom over microscopic distances on a time scale of the order of inverse jump frequencies; (iii) *the influence of impurities* on the particle motion is extremely important for muons which are heavily trapped by defects in the ppm range. But also the motional characteristic of hydrogen is severely affected.

The present review intends to address these three main features in an exemplary way. Recent theoretical and experimental work on muon and H diffusion will be surveyed. Thereby *it is the aim to elucidate the main developments* rather than to give a full account of all theoretical and experimental work which only as a consequence of the huge amount of recent publications would be a hopeless undertaking.

Other than the diffusive motions of heavy atoms, which in general are well described by classical diffusion theories, or the transport properties of electrons, which at least in metals propagate in band states and are of pure quantum nature, the diffusion of light interstitials comprises aspects of both: while bandlike transport of electrons is limited by scattering processes on impurities and phonons, heavy atoms move by thermal activation and pass continuously across saddle-point configurations. At not too low temperatures thermal activation also plays an essential role for the diffusive motion of light interstitials. However, their small mass leads to a finite overlap of the particle wave function at adjacent sites and tunneling processes through the potential barrier become possible. The combination of thermal activation, which is needed to overcome the lattice relaxation around the interstitial, and tunneling is subject of the small polaron theory and will be outlined in Sect.2.1-2.4. In particular, recent extensions of the theory beyond the so-called Condon approximation, where the dependence of the tunneling matrix element on the lattice coordinates is explicitly considered, and the modifications of the theory for disordered materials are discussed in some detail. At low temperatures direct tunneling of the interstitial together with its lattice deformation is also possible and small-polaron bands are conceivable. The transport properties in such small-polaron bands are outlined in Sect.2.5. Special emphasis is dedicated to disorder-induced localization phenomena in such a band state.

As became evident experimentally, diffusion and trapping phenomena on impurities or defects are intimately related for the muon. Therefore, Sect.2.6 includes trapping events in the diffusion process. They are described in the framework of a two-state model which considers diffusion in the presence of traps as a sequence of transitions between a propagating and a trapped state characterized by their respective lifetimes τ_1 and τ_0.

Finally the conditions for small-polaron formation are examined in Sect.2.7. While the small-polaron state comprising the light interstitial and its lattice deformation always constitutes the ground state for the particles considered here, under certain conditions barriers to self-trapping appear to be possible and could affect the muon behavior within the experimental observation time.

Experimentally, the nature of the elementary diffusive step can be unraveled from an investigation of the temperature and isotope dependence of the particle jump rate. Besides the well-documented results for H(D,T) diffusion which are reviewed in Sect.4.1, in particular muon diffusion experiments revealed new insights into light-interstitial diffusion under extreme conditions. After a brief description of the method, they are surveyed in Sect.3. Thereby this review restricts itself to experiments on nonmagnetic materials. Firstly, results on the muon lattice location in various fcc and bcc metals are presented. In all instances the muon interstitial sites resemble closely the sites found for H, demonstrating, thereby, the similarity of the muon and the H host interaction. In Sect.3.5 the early μSR (muon spin rotation) results on Cu are recalled, which clearly reveal subbarrier diffusion for the muon while the diffusion mechanism for H is closer to an over barrier jump process. Thereafter, taking Nb as an example, Sect.3.5 traces how the understanding of muon behavior in metals developed. In particular, the crucial importance of minute amounts of impurities comes into focus. Not only they are responsible for diffusion-controlled muon trapping, but in Nb they also appear to catalyse small-polaron formation itself. Finally, in Sect.3.6, muon diffusion studies on Al and dilute $\underline{Al}Me_x$ alloys are presented. These most detailed studies on muon diffusion established the temperature dependence of the muon diffusion coefficient D^{μ} over more than 3 orders of magnitude in temperature. Around 3 K D^{μ} passes through a minimum. For $T < 3$ K "coherent" band-like diffusion becomes prominent which increases with decreasing T, while for higher T phonon-assisted motion takes over and causes a growing D^{μ} with growing temperature.

The space and time evolution of the hydrogen diffusion process which is probed best by quasielastic neutron scattering (QNS) is the main subject of Chap.4. The results on H diffusion in metal-hydrogen systems are discussed in the order of their complexity. Commencing with diffusion of dilute H hopping between nearest-neighbor sites in a Bravais lattice, the concepts of QNS are outlined on the example of PdH_x which appears to be a physical realization of this most simple case. The review continues with a description of H diffusion in a non-Bravais sublattice, as realized, e.g., in bcc Nb, where at higher temperatures strong deviations from nearest-neighbor jump models are also evident. Other than in Pd, sequences of H jumps are highly correlated and are described best by a two-state model comprising self-trapped and propagating modes. As a further complication Sect.4.2.3 takes trapping on impurities into account. Results on N-doped Nb are well described by the two-state trapping model mentioned above. In particular explicit results on the trapping rate τ_1^{-1} and the escape rate from a trap τ_0^{-1} were obtained.

Section 4.2.4 deals with the influence of finite H concentration on the H jump process. The increasing H concentration introduces two new features: (i) blocking effects due to the presence of other protons, and (ii) backward correlations between successive jumps. A description of hydrogen diffusion in intermetallic hydrides, which are of great interest with respect to their application as H storage materials, requires a combination of the features originating from H trapping and those from finite H concentrations. Due to their complicated crystallographic structures and their off-stoichiometry, structural traps are present in hydrides which modify the H diffusion properties in addition to the mutual blocking of H atoms. Recent results on $Ti_{1.2}Mn_{1.8}H_3$ demonstrate the application of these ideas.

Section 4.3 is concerned with hydrogen vibrations and their relation to diffusive motion. From the local oscillations of the H atom against its metal neighbors, the strength of the metal-hydrogen interaction and therewith the hydrogen potential is obtained which is an important input parameter for all diffusion theories. New results on the local H vibrations in bcc refractory metals are presented. Thereby special emphasis is imposed on anharmonicities and isotope effects. The existence of second harmonics well above the activation energies for H diffusion obviously shows that H diffusion cannot be envisaged as jump diffusion over potential barriers provided by the vibrational potential. Coupling to the acoustic phonons of the host lattice is seen in the so-called band modes. Resonant-like enhancements of the H amplitude for certain modes may be a signature of lattice-activated diffusion processes in these materials.

Finally Sect.4.4 deals with the dynamics of trapped protons. A large variety of H motions associated, e.g., with the O(N)-H complexes in Nb, could be identified. Internal friction reveals orientational relaxation processes of O(N)-H pairs which exhibit significant deviations from an Arrhenius-like T dependence. QNS yields additional fast local jump processes of the trapped proton. At low T, H tunneling processes become evident. The local vibrations of H trapped at interstitial or substitutional impurities resemble closely the vibrations in the pure lattice and allow an assignment of the H sites in the trapped state.

2. Transport Theory of Light Interstitials in Metals

2.1 Basic Considerations

The transport of light interstitials in metals is commonly described in the framework of the small polaron theory which was developed originally by HOLSTEIN /2.1,2/ and YAMASHITA and KUROSAWA /2.3/ in order to understand electronic hopping conduction in insulators and then applied to the problem of light interstitials in metals by FLYNN and STONEHAM /2.4/. The light interstitial is considered to interact with the surrounding metal atoms and to displace them such as to provide a potential "well" for the particle. For a sufficiently deep well the latter will occupy a bound, self-trapped state, unable to move unless accompanied by the well. The entity particle and distortion cloud together is called "small polaron".

Light interstitials, like the proton or the muon after thermalization, are screened by the conduction electrons of the metal. There remains a strong residual interaction with the host atoms mediated by the screening cloud. This interaction is often described in terms of lattice theory /2.5/, where the short-range interaction is modeled by so-called Kanzaki forces $\underline{\psi}^m$ to the surrounding host atoms designated by the indices m. The strength of the elastic interaction between the light interstitial and the host lattice can be parametrized in terms of the double force tensor $P_{\mu\nu}$ which is the first moment of the Kanzaki forces,

$$P_{\mu\nu} = - \sum_m \psi_\mu^m R_\nu^m \quad , \tag{2.1}$$

where R_ν^m is the νth component of a vector from the interstitial to the host atom m. In a cubic system the trace of $\underline{\underline{P}}$ is related to the volume expansion created by the lattice relaxation around a self-trapped particle by

$$\Delta V = \frac{Tr\{P\}}{C_{11} + 2C_{12}} \quad , \tag{2.2}$$

where C_{11} and C_{12} are elastic constants of the host metal. The elastic interaction favors self-trapping and small-polaron formation. For an isotropic elastic continuum the energy gain as a result of the lattice relaxation around the interstitial amounts to

$$\Delta U = - \frac{1}{2\Omega} \frac{P^2}{C_{11}} \quad , \tag{2.3}$$

where P is a diagonal element of $\underline{\underline{P}}$ and Ω is the volume of the primitive cell. Table 2.1 presents some typical values of P and ΔU for H in some selected bcc and fcc metals. Figure 2.1 displays schematically the relaxed lattice configuration around a self-trapped interstitial. With respect to the undisturbed lattice the potential is lowered by ΔU. The relative energy difference to the adjacent site δU can be calculated from the elastic interaction energy with an identical particle at this site. For protons in Nb HORNER and WAGNER /2.7/ have computed δU in the framework of elastic lattice theory and found δU = 109 meV.

Table 2.1. Double force tensors P and energy gain by lattice relaxation ΔU for H in selected bcc and fcc materials

Metal	P [eV][a]	ΔU [eV][b]
Ta	3.3	-0.19
Nb	3.3	-0.20
Pd	3.5	-0.30

[a] The double force tensors are obtained from the relative volume changes $\Delta V/\Omega$ caused by H in the different metals /2.6/.

[b] ΔU is calculated from (2.3).

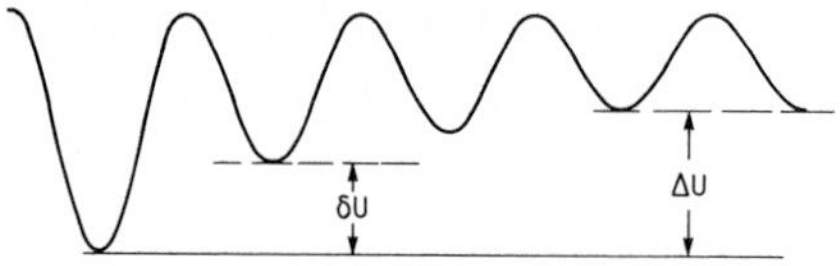

Fig.2.1. Schematic sketch of the interstitial potential in the neighborhood of a self-trapped interstitial

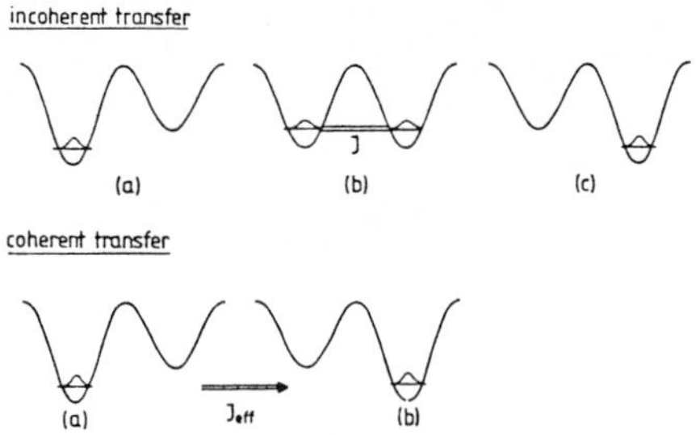

Fig.2.2. Quantum diffusion processes for light interstitials; *incoherent process:* lattice fluctuation equalizes adjacent potential wells and tunneling occurs (J); *coherent process:* interstitial tunnels together with its deformation cloud (J_{eff}) without the aid of phonons

Classically, the transition rate to a neighbor position is given by a trial frequency which is related to the vibrational frequency in the potential well ν and a Boltzmann factor containing the saddle point energy U_a /2.8/

$$\Gamma_{ab} \sim \nu \exp(-U_a/kT) \quad . \tag{2.4}$$

For the derivation of (2.4) a continuous motion across the saddle point is assumed. For light interstitials with their widely spaced vibrational levels and their ability to tunnel through potential barriers this classical picture is not adequate and small-polaron theory has to be applied, which considers the quantum-mechanical transition rate between different sites. As it is evident from Fig.2.1 the potential levels between adjacent sites are energetically separated by δU and direct tunneling is inhibited. There remain two general possibilities (Fig.2.2).

(i) Lattice fluctuations may equalize the potentials between adjacent sites and form a so-called coincidence configuration during which the interstitial may tunnel between both sites. Then, the tunneling matrix element J is given by the overlap of the particle wave functions at different sites in the coincidence configuration which is often approximated by the rigid nonrelaxed lattice. Such a mechanism is a multiphonon process which at high temperatures follows an Arrhenius relation like classical diffusion, though, as we shall see, the parameters involved have a different meaning. At lower temperature hopping of the particle dressed by its distortion cloud becomes important. The dressing reduces the overlap to an effective value J_{eff} which takes into account that not only the particle but also the distortion cloud has to tunnel. Both jump processes are phonon assisted. In general, due to energy conservation the lowest-order process is a two-phonon mechanism. Since there is no phase relation between the particle wave function before and after the jump, these transfers are also termed "incoherent".

(ii) Under appropriate conditions, at low temperatures, also direct tunneling processes of the "dressed particle" appear to be possible which do not involve the aid of phonons. A small-polaron band is formed consisting of Bloch-like states, where the propagation is limited by various scattering mechanisms. This bandlike transport is also named coherent diffusion, since phase relations are maintained. Similarities to narrow-band electronic conductors are anticipated.

In order to evaluate actual jump or scattering rates normally a series of simplifying assumptions are introduced into small-polaron theory:

(i) Adiabatic decoupling of particle and host (the interstitial follows the host atoms adiabatically). This assumption is justified as long as the local mode frequencies for H(D) or μ^+ are much larger than the Debye frequency of the host.

(ii) Linear coupling between particle and host (interaction energies are linear in the displacement of the host; as a consequence the atoms of the host only are displaced. Their vibrational properties are not altered).

(iii) Localized particles are assumed to constitute the unperturbed ground state. This is justified, since the energy gain by lattice relaxation ΔU is in general large compared to the tunneling matrix element J. The influence of J which favors a translational invariant band state is considered in a perturbational treatment.

(iv) Harmonic approximation of the lattice.

(v) J is independent of the displacements caused by the phonons. This is called Condon approximation.

Though some progress has been reported in overcoming the adiabatic approximation and the linear coupling approach /2.9-11/, we shall maintain assumptions (i)-(iv) for the sake of clarity and since no qualitative changes seem to appear in removing (i) and (ii). The Condon approximation on the other hand has proved itself a severe restriction and we shall investigate the consequences of its removal.

Within the Condon approximation only those phonons are considered which equilibrate adjacent potential wells, while phonons affecting the potential barrier are not taken into account. Changes of the barrier, however, severely influence the tunneling matrix element J and thus the probability that a transfer takes place. Two limiting temperature regimes have to be distinguished:

(i) At low T, where mainly lowest energy coincidences occur, J in general is small and rate determining. This regime is called nonadiabatic since the small polaron does not follow all the offered transfer possibilities.

(ii) At higher T higher energy coincidences occur. Here J is large and the particle exploits all coincidence events. The corresponding temperature regime, therefore, is termed adiabatic.

2.2 Phonon-Assisted Motion – Classical Small-Polaron Theory

FLYNN and STONEHAM adopted the ideas of small-polaron motion for the diffusion of light interstitials in metals /2.4/. Disregarding any delocalization, they calculated the quantum-mechanical transition rates between different interstitial positions. Since band states rather than localized wave functions are the eigenstates of the small polaron Hamiltonian H, it induces transitions between localized states at positions p and p' which give the particle motion we are interested in. The transition rates between specific sites $w_{pp'}$ are obtained from a golden-rule calculation

$$w_{pp'} = \frac{2\pi}{\hbar}|\langle p,\alpha|H|\alpha',p'\rangle|^2 \ \delta(E_{p\alpha}-E_{p'\alpha'}) \quad , \qquad (2.5)$$

where $|\alpha,p\rangle$ describes a localized state at the position p and α contains all the other quantum numbers of the system; $E_{p\alpha}$ is the energy corresponding to state $|p,\alpha\rangle$. In general a transition from p to p' is possible involving different sets α and α'. Since we are interested only in the total rate, an average over all initial states $\langle p,\alpha|$ and a sum over all final states $|\alpha',p'\rangle$ has to be carried out. The details of the calculation are rather lengthy and exhausting. Therefore, we omit them and present only some of the results which consecutively we shall try to make physically plausible.

For temperatures in the order of or above the Debye temperature θ_D the small-polaron transition rate in Debye approximation is given by

$$\Gamma_{pp'} = \left(\frac{\pi}{4\hbar^2 E_a kT}\right)^{1/2} J^2 \exp(-E_a/kT) \quad , \qquad (2.6)$$

where E_a is the energy necessary to create the coincidence situation. Further, E_a is related to the energy shift between adjacent sites δU by $E_a = (1/4)\delta U$ and is not connected with the height of the potential barrier between two interstitial sites which determines the activation energy for classical diffusion.

Qualitatively this result can be understood as follows /2.12/. Assume that for a certain displacement q_0 a coincidence event takes place. For a classical oscillator the probability $P(q_0)$ for such a displacement is given asymptotically by

$$P(q_0) \sim T^{-1/2}\exp[-w(q_0)/kT] \quad ,$$

with $w(q_0) = M\,\omega^2 q_0^2/2 = E_a$, where ω and M are the corresponding frequency and mass. According to the golden rule, the transition rate is also proportional to J^2, and we obtain the desired rate $\Gamma_{pp'}$ as a product of the occurrence probability and J^2

$$\Gamma_{pp'} \sim J^2 T^{-1/2}\exp[-w(q_0)/kT]$$

which besides numerical factors resembles the result of the detailed calculation.

For low temperatures the transition rate is a power of the temperature. For an isotropic solid the jump rate becomes

$$\Gamma_{pp'} \cong 57600\,\pi\,\omega_D \frac{J^2}{(\hbar\omega_D)^4} E_a^2\, e^{-5E_a/\hbar\omega_D} (T/\theta_D)^7 \tag{2.9a}$$

$$\cong 57600\,\frac{\pi}{\hbar} \frac{E_a^2}{(\hbar\omega_D)^3} (J_{eff})^2 (T/\theta_D)^7 \tag{2.9b}$$

with $\hbar\omega_D = k\theta_D$. The high temperature power T^7 is a consequence of a two-phonon process.[1] In ideal lattices a small polaron cannot emit or absorb a single phonon because of energy and momentum conservation, but it can virtually absorb a phonon, tunnel to the neighboring site and emit it again. The responsible tunneling matrix element J_{eff} is reduced with respect to J, since at low T coincidences which equalize initial and final state do not occur and the whole polaron (particle and distortion cloud) has to tunnel.

Applying again an isotropic Debye approximation, STONEHAM has also given a closed expression for the jump rate $\Gamma_{pp'}$ which is valid over the entire temperature range /2.14/. It has the form of an integral and can, e.g., be used to investigate the limits of the high- and low-T approximations. Their validity was found to depend on the ratio $E_a/\hbar\omega_D$. For reasonable values the Arrhenius law is valid above θ_D whereas the T^7 law holds below about 10% of θ_D.

The theoretical calculation of the parameters J and E_a from first principles is extremely difficult. An estimation of J requires exact knowledge of the potential energy curve for the particle along its diffusive path in the lattice. A computa-

[1] The exact power depends on the model and on lattice properties, e.g. for several inequivalent sites within the unit cell FUJII has obtained a T^3 behavior of the two-phonon transition rate /2.13/.

tion of E_a needs information on the forces between the interstitial and the surrounding atoms as well as on the forces between the host atoms. As already mentioned, for bcc Nb δU has been calculated using the phonon Green function and Kanzaki forces derived from the double force tensor of H in Nb. Taking $E_a = \delta U/4$ /2.15/ a value of 27 meV results. TEICHLER /2.16/ has estimated the small-polaron jump rate between octahedral sites in a fcc lattice using a model with two parameters: the Kanzaki force ψ from the H isotope to its nearest neighbors and one spring constant f representative for the lattice. For the activation energy he found

$$E_a = 0.258\ \psi^2/f \quad . \tag{2.10}$$

On the basis of a nonlinear screening approach TEICHLER /2.17/ has also evaluated the Kanzaki forces ψ for H isotopes in Cu. For the muon, e.g., he found $\psi a = 2$ eV, where a is the lattice constant. Inserting $f = a\, c_{44} = 2.95 \cdot 10^4$ dyn/cm this yields $E_a = 43$ meV. His calculations predict a strong isotope effect between the Kanzaki forces for the muon and those for the heavier isotopes. The corresponding activation energies are distinctly smaller and decrease slightly with increasing mass (E_a = 19; 17; 16 meV for the proton, the deuteron and the triton respectively).

Table 2.2. Estimates for the tunneling matrix elements J starting from a 1-d cos potential which reproduces the local H vibrations and the interstitial distances d

Metal	d [Å]	J_{μ^+} [meV]	J_H [meV]	J_D [meV]
V	1.07	69	0.13	$3.1 \cdot 10^{-3}$
Nb	1.17	43	$2.5 \cdot 10^{-2}$	$2.9 \cdot 10^{-4}$
Pd	2.74	0.20	$6.5 \cdot 10^{-9}$	$1.6 \cdot 10^{-13}$
Cu	2.55	$5.9 \cdot 10^{-3}$	$3.7 \cdot 10^{-14}$	$4.5 \cdot 10^{-21}$

An estimation of J is far more intricate, since J depends exponentially on the interstitial potential, and subtle changes can have a heavy impact on the value of the overlap integral. In order to obtain an order of magnitude estimate KEHR /2.18/ assumed a one-dimensional periodic potential for the H isotopes and calculated the energy splitting between the symmetric and antisymmetric ground state components. The potential parameters were chosen such as to reproduce the local vibrations and the interstitial distances. Table 2.2 lists some of his results. Included are also estimates for Cu which have been obtained in his spirit using a value of $\hbar\omega_H = 120$ meV as evaluated by TEICHLER /2.17/ (experimental data are not available). Though the values for J are only rough estimates, they clearly show the crucial influence of the isotope mass as well as the tendency that J decreases with increasing jump distances d.

Finally, the dependence of the jump rate on the isotope mass, which results from this "classical" small-polaron theory, is summarized as follows: the weakly temperature-dependent prefactor in the expressions for $\Gamma_{pp'}$ is proportional to J^2 and extremely sensitive to the isotope mass as can be seen from Table 2.2. The influence of the isotope mass on the activation energy in the Arrhenius relation at high temperatures is directly proportional to the square of the Kanzaki forces. For the three isotopes H, D and T, the double force tensors generally depend only weakly on the isotope masses, e.g., for Nb no isotope effect on P could be detected, for Ta $P_H > P_D$ was observed /2.6/. Thus, for the small-polaron jump rate we expect a strongly isotope-dependent prefactor and an activation energy only weakly affected by the isotope mass.

2.3 Small-Polaron Hopping Beyond the Condon Approximation

So far we have assumed that the overlap integral J is independent of the actual lattice configuration Q. Already FLYNN and STONEHAM /2.4/ pointed out that for certain lattices and jump geometries like jumps between octahedral sites in fcc lattices this assumption constitutes a severe restriction and should be removed. In order to understand its influence, we start with an expression /2.15,18/ for the activation energy in terms of normal coordinates $q^{p}_{\underline{K}\lambda}$ and $q^{p'}_{\underline{K},\lambda}$ of the displacements around the self-trapped particle on site p and p'

$$E_a = \frac{1}{8} \sum_{\underline{K},\lambda} M\, \omega^2_{\underline{K}\lambda}\, \delta q_{\underline{K}\lambda}\, \delta q^*_{\underline{K},\lambda} \qquad (2.11)$$

with

$$\delta q_{\underline{K}\lambda} = q^{p}_{\underline{K}\lambda} - q^{p'}_{\underline{K}\lambda} \quad ,$$

where $\underline{K}$ denotes the wave vector and λ assigns the phonon mode. From (2.11) it is evident that only modes with components parallel to $\delta q_{\underline{K}\lambda}$ enter into E_a. In particular, for isotropic cubic dilatations only modes which are antisymmetric with respect to the midpoint of the jump path contribute to E_a. The symmetric modes which do not change the relative shift of the interstitial potential δU and consequently are not represented in (2.11), however, may severely influence the potential barrier and thus the tunneling matrix element J. This is demonstrated in Fig.2.3, where a (0,0,1) plane at z = a/2 in an fcc lattice is shown. The dashed line indicates the

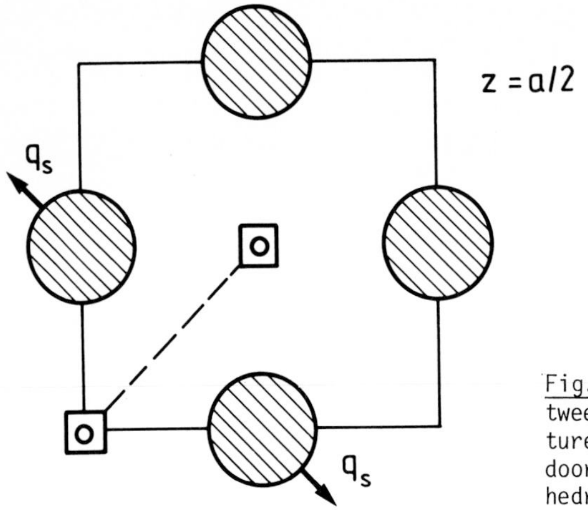

Fig.2.3. Lattice activation for jumps between two octahedral sites in an fcc structure. The symmetric mode q_s "opens the door" for the jump between the two octahedral sites (O sites)

the jump path between two octahedral sites and the arrows symbolize a symmetric mode q_s, which "opens the door" for the interstitial and should strongly influence J. For such a situation, where J is expected to be very sensitive to q_s, the Condon approximation has to be dropped. In a first simple approach FLYNN and STONEHAM /2.4/ suggested a tunneling matrix element $J = 0$ for q_s smaller than a critical displacement q_s^c and $J = J_0$ for $q_s > q_s^c$. In the high-temperature limit this approach leads to

$$\Gamma_{pp'} = \frac{1}{4\hbar} \frac{J_0^2}{\sqrt{E_a E_s}} \exp\left[-(E_a+E_s)/kT\right] \quad , \tag{2.12}$$

where E_s is the elastic energy necessary to create the critical distortion q_s^c. For a qualitative derivation of (2.12) we refer again to our occurrence probability approach /2.12/. Now we have to ask for the probability $\tilde{P}(q_s^c)$ that a displacement q_s exceeds q_s^c. For a classical oscillator $\tilde{P}(q_s^c)$ is given asymptotically by

$$\tilde{P}(q_s^c) \cong T^{1/2} \exp(-E_s/kT) \quad . \tag{2.13}$$

The jump rate is then the product of the probability to create a distortion larger than q_s^c, the probability to produce a coincidence event for the tunneling process (2.7) and J^2

$$\Gamma_{pp'} \sim J^2 T^{1/2} T^{-1/2} \exp\left[-(E_a+E_s)/kT\right] \quad , \tag{2.14}$$

in agreement with (2.12).

FLYNN and STONEHAM /2.4/ argue that whenever atoms are straddling the jump path like for jumps between octahecral sites in fcc metals, the Condon approximation ceases to be valid and E_s contributes an important part to the activation energy. As KEHR /2.18/ pointed out, this is also true for jumps between tetrahedral sites (T sites) in bcc metals, though the effect may be smaller than in fcc metals.

An important step forward was achieved by EMIN et al. /2.19/ who studied numerically the dependence of J on the lattice coordinates. They assumed the vibrational motions of the host as being classical and described the jump processes in an occurrence probability approach. As we have seen already in our qualitative explanation of $\Gamma_{pp'}$ in the high-T limits, in this approach the expression for the jump rate is separated into two contributions, namely: (i) a probability $P_{pp'}$ for the occurrence of coincidence events, where the energy level associated with a light particle occupying site p momentarily coincides with the energy at an adjacent site p', and (ii) a probability $R_{pp'}$ that at a given coincidence the particle actually negotiates a hop. Two limiting cases are obvious: (i) for small tunneling matrix elements $R_{pp'}$ will be much smaller than one and the chance that a coincidence situation actually leads to a jump is small and strongly dependent on the magnitude of J. We are in the nonadiabatic regime; (ii) for large J the dependence of $R_{pp'}$ on J becomes progressively weaker and approaches unity in the so-called adiabatic regime. Here, the magnitude of J does not enter the jump rate any more.

EMIN et al. /2.19/ performed explicit calculations on a one-dimensional model using potential parameters such as to simulate the metal-metal and metal-hydrogen interaction in Nb. The coincidence events were described in terms of a single deformation parameter associated with symmetric displacements similar to the ideas on lattice activation put forward by FLYNN and STONEHAM /2.4/. Though the numerical results depend on the actual choice of parameters, some features are of general bearing. (i) The tunneling matrix elements are very sensitive to changes of the coincidence deformations. As an example the authors mention that an increase of the coincidence energy by only 0.03 eV increases J by a factor of 30; (ii) excited states within a coincidence situation also contribute to the small-polaron transfer.

The results of the model calculation for Nb are shown in Fig.2.4, where the calculated diffusion coefficients are displayed in an Arrhenius plot. The following features should be noted: (i) in contrast to the results of the theory within the Condon approximation, the high-temperature activation energy for H diffusion increases with increasing mass, reflecting the necessity for the heavier particle to search for higher energy coincidences in order to reach the adiabatic regime; (ii) the adiabatic character of the high-temperature diffusion behavior manifests itself in an isotope-independent prefactor ($R_{pp'}$ does not depend on J in the adiabatic re-

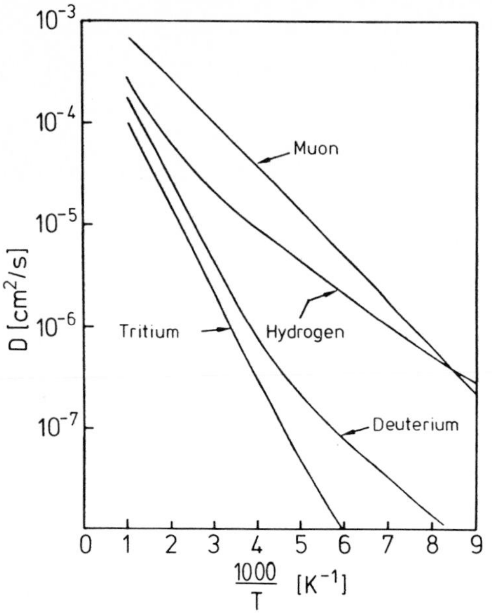

Fig.2.4. Calculated diffusion coefficient for H and its isotopes as well as for the positive muon in Nb /2.19/

gime); (iii) at lower temperatures the contribution of nonadiabatic transfers which require a lower activation energy increases. This results in an apparent change of the activation energy for diffusion which with increasing isotope mass occurs at decreasing temperatures. In the nonadiabatic regime the predictions of ordinary small-polaron theory are retained; (iv) for the muon already the lowest energy coincidence leads to large transfer matrix elements. Thus, the muon shows an activated behavior according to this energy over the whole temperature range.

Finally, TEICHLER /2.20,21/ performed a rigorous microscopic treatment beyond the Condon approximation in the nonadiabatic limit. He dropped the assumption of classical host motion implicit in the occurrence probability approach and treated the host vibrations quantum-mechanically. As a consequence the separation into coincidence and transfer probabilities is not possible any longer and the transition rate is determined by correlated averages over all phonons which cause a fluctuational reduction of the barrier *and* fluctuations of the small-polaron levels. For special cases, explicit results on the temperature dependence of $J_{pp'}$ were derived and are presented in App.1. The essential feature is a distinct transition from nonactivated to activated behavior for large correlations between barrier and potential well fluctuations. Similarly to Emin's considerations, such a transition would manifest itself in an apparent change of the activation energy for small-polaron diffusion.

2.4 The One-Phonon Process

In ideal crystals jump processes between equivalent sites, which occur by the emission or absorption of *one* phonon only, in general violate the requirement of energy conservation and are forbidden. As a consequence the prevalent low-temperature transfer involves two phonons which leads to the predicted high-temperature powers of the jump rate (2.9). In a real lattice with impurities, dislocations, grain boundaries, etc., the translational invariance is violated and the interstitial sites are no longer equivalent. Apart from the fact that imperfections create trapping regions, treated in Sect.2.6, they cause energy differences ΔE between adjacent sites even far away from the source. Under these circumstances a particle may hop already under emission or absorption of one phonon which then compensates for this energy difference ΔE.

Phonon-assisted tunneling in systems with energy shifts ΔE between adjacent sites was first treated by SUSSMANN /2.23/, who neglected small polaron effects. Later on, his ideas were applied to describe the influence of lattice strains on the tunneling process involved in the reorientation of dipolar defects /2.24/. Recently, TEICHLER and SEEGER /2.25/ investigated the occurrence of one-phonon processes in the small-polaron framework within the Condon approximation. According to their calculations, a one-phonon process may occur (i) either in disturbed systems or (ii) between crystallographically inequivalent sites. In an isotropic Debye approximation the one-phonon jump rate between crystallographically equivalent but energetically distorted sites becomes

$$\Gamma^{1Ph} = J_{eff}^2 \frac{d^2 P^2 \Delta E^2}{12\pi\rho\hbar^6 c^7} \frac{\Delta E}{\exp(\Delta E/kT)-1} \quad , \qquad (2.15)$$

where d is the jump distance, ρ the mass density of the host and c the velocity of sound. For phonon emission ΔE has to be replaced by $|\Delta E|$ and the rate to be multiplied by $\exp(|\Delta E|/kT)$. For $kT >> \Delta E$ we obtain

$$\Gamma^{1Ph} = \frac{J_{eff}^2 d^2 P^2}{12\pi\rho\hbar^6 c^7} (\Delta E)^2 kT \quad , \qquad (2.16)$$

i.e., the rate is proportional to T and depends explicitly on the magnitude of the disturbance ΔE. For purified materials the most important remaining distortion stems from the long-range elastic interaction with impurities $[V(r) \sim 1/r^3]$. The relative energy shifts between adjacent sites are proportional to the derivative of the potential yielding $\Delta E \sim n_t^{4/3}$ where n_t is the impurity concentration. Thus with increasing purity of a material the one-phonon rate should vanish as $n_t^{8/3}$.

For a very pure material one-phonon processes remain possible only between inequivalent sites. Then, again in isotropic Debye approximation, the rate

$$\Gamma^{1Ph} = \frac{J_{eff}^2}{2\pi\rho c^5\hbar^4}\left[\frac{d^2 Tr\{P_i+P_f\}^2}{216\hbar^2 c^2}(\Delta\varepsilon)^2 + P_a^2\right]\frac{\Delta\varepsilon}{\exp(\Delta\varepsilon/kT)-1} \quad (2.17)$$

depends as before on the energy difference now between the inequivalent sites $\Delta\varepsilon$ and in addition on an asymmetric combination of the double force tensors in the initial and final sites $P_a = [Tr\{P_i - P_f\}]/2$. In this approximation P_a contributes to the jump rate only, if the magnitude of P changes during the jumps. This may occur, for example, if the particle diffuses between sites of different symmetry like O-T-O-T- sequences in fcc metals. We note that if the orientational averaging leading to (2.17) is carried beyond an isotropic Debye approximation, contributions from terms of the form $Tr\{P_i - P_f\}^2$ appear in (2.17). Then orientational changes of an anisotropic double force tensor during a jump also give rise to one-phonon transfers /2.25,26/. The latter case may occur in bcc metals, where the interstitial sites have tetragonal symmetry and jumps between nearest neighbors involve changes of the tetragonal axis. However, in spite of the tetragonal symmetry for H in bcc metals, isotropic or nearly isotropic double force tensors were observed /2.6/.

Finally, TEICHLER /2.26/ considered the influence of the Condon approximation on one-phonon transfers and found a transition rate which for certain symmetry requirements does not depend directly on the disturbances ΔE (App.2). Explicit evaluations, however, on, e.g., the order of magnitude of this contribution, are not available and its significance has still to be shown.

2.5 Coherent Motion

2.5.1 Small-Polaron Bands

In this section we shall consider direct tunneling of H isotopes including the surrounding distortion cloud. At low temperature and in sufficiently pure materials a self-trapped particle does not necessarily mean a localized particle. The small polaron may tunnel to neighboring interstitial sites and develop a narrow small-polaron band consisting of Bloch-like states. Contrary to small-polaron formation which is supported by a gain of elastic interaction energy (Sects.2.1,7) localization in a small-polaron band does not lead to further minimization of energy and other localization mechanisms have to be considered. Two possibilities, (i) strain localization and (ii) localization through phonon scattering, will be discussed later on.

The bandwidth of the small-polaron band is given by $z \cdot J_{eff}$, where z is the coordination number of the lattice and J_{eff} comprises the transfer element J of the particle in the rigid lattice and the transfer of the relaxed lattice configuration. For the case of strong coupling $|\Delta U| >> J$, J_{eff} is given by

$$J_{eff} = J\exp[-S(T)] \quad . \tag{2.18}$$

For harmonic lattices and linear coupling, which is generally assumed for small polarons, S(T) is determined by the Kanzaki forces ψ_{μ}^{m} and the lattice force constants. For a first estimate an isotropic Debye approximation suffices /2.18/ and yields

$$S(0) = \frac{5E_a}{2\hbar\omega_D} \quad . \tag{2.19}$$

As examples we give the S(0) values for the muon in Cu and Al: $S(0)^{Cu} = 6.34$ and $S(0)^{Al} = 2.16$ which were calculated from experimental activation energies obtained from diffusion measurements in these substances. They reduce J at T = 0 by a factor of $1.8 \cdot 10^{-3}$ or 0.11 respectively. The temperature correction of S(T) for small T is proportional to $(kT/\hbar\omega_D)^4$ /2.14/ and is of no importance at low T. For muons in Cu and Al, experimental values for J from high-T diffusion data of $J^{Cu} = 18$ µeV /2.16/ and $J^{Al} \cong 2$ meV /2.27/ have been deduced. For the small-polaron bandwidth these values yield $zJ_{eff}^{Cu} \cong 1$ µeV and $zJ_{eff}^{Al} \cong 2$ meV.

In real lattices such small-polaron bands are not realistic even at T = 0, since impurities and defects which are always present provide isolated localized low-energy states which would trap the particle eventually. In the case of the muon, which starts under nonequilibrium conditions and has only a limited lifetime, those trapping regions may not be reached in a material of sufficient purity. Under these circumstances it is necessary to discuss whether a formation of extended wave packets is possible at all in a crystal with defects far away from the traps. For this purpose we utilize ANDERSON's treatment /2.28/ of electron localization in randomly disturbed materials.

In his classical paper on the "Absence of Diffusion in Certain Random Lattices" Anderson argued that for sufficiently disordered systems the electron eigenstates localize. In particular, he found that for a distribution of random energy shifts ΔE in between the different lattice sites there exists a critical width Γ_c of the distribution above which localization occurs. It is essentially given by the bandwidth zJ_{eff}. As a prerequisite for coherent transport KAGAN and KLINGER /2.29/ demand crystal regions where the mean energy difference ΔE between adjacent sites is smaller than the tunneling matrix element J_{eff} between those sites. Qualitatively their arguments agree with those of Anderson. For the muon and the proton with masses much heavier than the electron mass the corresponding bandwidth is much smaller than the electronic one and even very weak disturbances could suffice to fulfill the Anderson criterion.

In the most purified, isotopically pure metals like Nb or Al the most important remaining disturbances far away from the actual impurity sites originate from the long-range elastic interaction between substitutional or interstitial impurities and the muon. We estimate the corresponding energy shifts ΔE from elasticity theory. For an approximation we employ the first term of a series expansion for the interaction of isotropic impurities in an anisotropic cubic material first given by ESHELBY /2.30/:

$$E_{int} = -\frac{1}{r^3}\frac{15}{8\pi} d_a \left(\frac{\bar{C}_{11}+2\bar{C}_{12}}{3\bar{C}_{11}}\right)^2 \Delta V^{\mu}_{H} \Delta V^{imp} \left(\frac{2}{5} - \sum_j \rho_j^4\right) \quad , \qquad (2.20)$$

where $d_a = C_{11}-C_{12}-2C_{44}$ is the anisotropy of the lattice, $\bar{C}_{12} = C_{12}+d_a/5$ and $\bar{C}_{11} = \bar{C}_{12}+2C_{44}+(2/5)d_a$ are averaged elastic constants, ΔV^{μ}_{H} and ΔV^{imp} are the volume expansion induced by the light interstitial and the impurity respectively, ρ_j are the direction cosines with respect to the cubic axis. We demonstrate the order of magnitude of the disturbances due to the elastic muon impurity interaction for the example of muons in Al with different amounts of substitutional Mn impurities. For this estimate we omit the angular dependence of the interaction and use $\Delta V^{\mu} = 2.86$ Å^3 which has been found for hydrogen in fcc metals /2.31/. Inserting the elastic constants for Al, (2.20) yields

$$E_{int} = -\frac{5.68\cdot 10^{-2}}{r^3} \Delta V^{imp} \; [eV] \quad . \qquad (2.21)$$

The energy change between adjacent sites in a distance d is then given by

$$\Delta E = \left(\frac{dE_{int}}{dr}\right) d = \frac{0.17\, d}{r^4} \Delta V^{imp} \; [eV] \quad . \qquad (2.22)$$

An estimate for the regions far away from the impurities is obtained by calculating ΔE at a point midway between them. Taking $\Delta V^{Mn} = -7.4$ Å^3 which can be deduced from the lattice dilatation of Mn in Al /2.32/, Mn at concentrations of 10, 100 and 1000 ppm leads to disturbances ΔE of 0.2, 5 and 100 µeV respectively. In view of the large bandwidth of $\sim$ 2 meV, these impurity concentrations, except the largest one, may not suffice to localize the muon. In contrast to this situation, where the muon small-polaron band is expected to be rather insensitive toward impurities, stands the case of Cu, where the bandwidth was estimated to be in the order of 1 μeV. Here already very small impurity concentrations should be able to prevent delocalization even at T = 0.

From the above considerations it is evident that hydrogen at any feasible dilution will always create random strain fields by itself such as to localize at all temperatures.

Above T = 0 dynamic destruction of small-polaron band states, e.g., by phonon scattering or possibly electron scattering, also has to be considered. These scattering processes are intimately related to small-polaron transport and are treated in the next section on coherent diffusion.

2.5.2 Bandlike Propagation

For a delocalized small polaron above T = 0 a finite mean free path ℓ will originate from scattering processes even in ideal crystals. A diffusion coefficient for band propagation is associated with these scattering events which reads, apart from numerical factors /2.1,2/,

$$D = \langle v^2 \tau \rangle_{av} \cong \overline{v^2} \tau \quad , \tag{2.23}$$

where v is the particle velocity and τ^{-1} is a transport scattering rate. The average in (2.23) has to be performed over all band states weighted with the appropriate lifetimes τ. For electrons in metals v would correspond to the Fermi velocity, for thermalized muons or single protons we have to distinguish two limiting cases

$$v \cong \frac{J_{eff} d}{\hbar} \qquad kT \gg zJ_{eff} \tag{2.24}$$

$$v \cong v_{th} = \sqrt{\frac{3kT}{m^*}} \qquad kT \ll zJ_{eff} \quad . \tag{2.25}$$

In the high-temperature limit all band states are populated and v is essentially given by the tunneling matrix element, while for low temperature the band states are populated thermally and we expect a thermal velocity. For muons in Al, where zJ_{eff} has been estimated to $\sim$ 2 meV, well below 10 K v should be the thermal velocity. For muons in Cu on the other hand, where $zJ_{eff} \cong$ 1 μeV, at all practical temperatures we are in the high-T limit and v should be independent at T. The mean free path of the small polaron or the coherence length of the wave packet is given by $\ell = v\tau$.

The main task of small-polaron theory now is to evaluate the transport scattering rate τ^{-1}. This problem was approached firstly by HOLSTEIN /2.1,2/, who assumed the lifetime of a band state to be equivalent to the inverse total phonon-assisted jump rate. Thus, according to him, lifetimes in localized states and in polaron bands are equal and determined by the same scattering events. The pertinent process for the reduction of the coherent transfer rate with increasing temperature, thereby, is the interaction of phonons with the effective tunneling matrix element $J_{eff} = Je^{-S(T)}$, which at low T reduces J^{eff} only slowly with temperature.

However, not all scattering events which may lead to the destruction of band states contribute to the transport scattering rate, e.g. fluctuations of the relative shifts of energy levels in neighboring wells due to interaction between the particle and phonons do inhibit coherent transfers but do not automatically create incoherent jumps. These intrawell scattering events were taken into account by KAGAN and KLINGER /2.29/, who developed a unified theory of small-polaron transport including coherent motion. They assumed from the outset that J_{eff} is the smallest parameter of the system and started from (2.23) taking $v \sim J_{eff}$. Then τ^{-1} was derived from indirect phonon scattering events, where the small polaron is virtually in an excited state before it emits again the initially absorbed phonon. For the temperature dependence of the scattering rate they found $\tau^{-1} \sim T^9$. Recently KEHR /2.33/ rederived the scattering rate now assuming direct phonon scattering. In addition he expressed the prefactor of the scattering rate in terms of elastic parameters including the change of the elastic constants ΔC by adding the interstitial impurity:

$$\frac{1}{\tau} = \frac{8!\zeta(9)}{4\pi^3} \frac{\Omega^2 d^2 \overline{\Delta C}^2}{\rho^2 c^3} \left(\frac{kT}{\hbar c}\right)^9 \tag{2.26}$$

$$\overline{\Delta C}^2 = \Delta C_{12}^2 + \frac{4}{3} \Delta C_{12} \Delta C_{44} + \frac{1}{5} \Delta C_{44}^2 \quad .$$

Inserting elastic data for Cu a rate $1/\tau = 1.8 \cdot 10^5\ T^9\ [K]\ s^{-1}$ has been estimated. The relation $\tau^{-1} \sim T^9$ implies for the diffusion coefficient $D \sim T^{-9}$ for $zJ_{eff} \ll kT$ and $D \sim T^{-8}$ for $zJ_{eff} \gg kT$. In the case of several inequivalent interstitial sites like tetrahedral sites in bcc metals, FUJII /2.13/ has found a T^7 law for the scattering rate which reduces the exponents for the diffusion coefficient for such metals accordingly.

Beyond the Condon approximation KAGAN and KLINGER /2.34/ discussed the "fluctuational preparation of the tunneling barrier" in a model where J depends exponentially on the fluctuations. Under optimum conditions this may give rise to an increase of the rate with increasing T. In general, however, the fluctuations depend moderately on T only and the strong temperature dependence of the scattering rate will mask all changes of J.

For $T \ll 1$ K the phonon scattering rates decrease fast towards zero and unreasonably high diffusion coefficients would result. Therefore, additional scattering processes have to be considered. As pointed out recently by KEHR et al. /2.27/, an interaction process neglected so far (see, however, ANDREEV and LIFSHITZ /2.35/) is provided by muon-electron scattering. The electron scattering rate should be proportional to the number of excited electrons, i.e., $\sim T/T_F$, where T_F is the Fermi temperature. Further, an electron can at most transfer its thermal energy kT to the muon. Consideration of energy and momentum conservation yields an addition factor T/T_F in the scattering rate /2.36/. Electron scattering on muons should be the dominant scattering process at low temperatures in pure crystals.

On the basis of a Thomas-Fermi approximation KEHR et al. /2.27/ estimate an electron scattering rate τ_{el}^{-1} of the order of $2 \cdot 10^9\ T^2\ s^{-1}$. The resulting diffusion coefficient should be proportional to T^{-2} or to T^{-1}, depending on whether zJ_{eff} is smaller or larger than kT and differs in one T power from the suggestion of /2.35/.

Another scattering mechanism is the scattering on isolated impurities. As in the case of conduction electrons, such a process becomes important if the other scattering is weak and the mean free path approaches the mean distances between impurities. The scattering rate of impurities is given by

$$\frac{1}{\tau_d} \cong \sigma_{scat}\, n_t v \quad , \tag{2.27}$$

where σ_{scat} is the impurity scattering cross section. Under these circumstances the diffusion coefficient becomes inversely proportional to the impurity concentration and proportional to T^0 in the high-T limit. For $kT \ll zJ_{eff}$ $D \sim T^{1/2} n_t^{-1}$ is expected. Applying (2.27) and assuming a cross section of 1 Å^2 for muons in Al, an impurity scattering rate of

$$\frac{1}{\tau_d} \cong 10^{+5}\ T^{1/2}\ [K]\ n_t\ [ppm]\ s^{-1} \tag{2.28}$$

can be estimated.

In analogy to the Matthiesen rule it is plausible to evaluate the total lifetime of a band state τ which enters (2.23) from the sum of the different scattering rates

$$1/\tau = 1/\tau_{ph} + 1/\tau_{el} + 1/\tau_d + \dots \tag{2.29}$$

2.5.3 Transition Between Coherent and Incoherent Diffusion

After we have treated coherent and incoherent diffusion which should prevail at low and high temperatures respectively, it is natural to ask for the transition temperature T_c between both regimes. This problem was addressed firstly by HOLSTEIN /2.2/. He came to the conclusion that the transition temperature is reached if the energy uncertainty associated with the lifetime of the band states becomes equal to the bandwidth. At this temperature the Bloch states are no longer meaningful and can be considered as washed out. Since he assumed that the same scattering processes are responsible for transport as well as for the lifetimes of the small-polaron bands, his criterion also means that at the transition temperature the hopping diffusion rate surpasses the coherent rate. This is shown schematically in Fig.2.5 where $J_{eff}(T)$ as well as the incoherent rate $\tau^{-1}(T)$ are plotted vs. the temperature. Thereby the reduction of the coherent transfer rate stems from the interaction of the phonons with the effective tunneling matrix element $J_{eff} = Je^{-S(T)}$ which at low T depends only weakly on the temperature and for temperatures of some tenth of θ_D falls off exponentially with T. For the transition temperature Holstein estimated $T \cong \theta_D/2$.

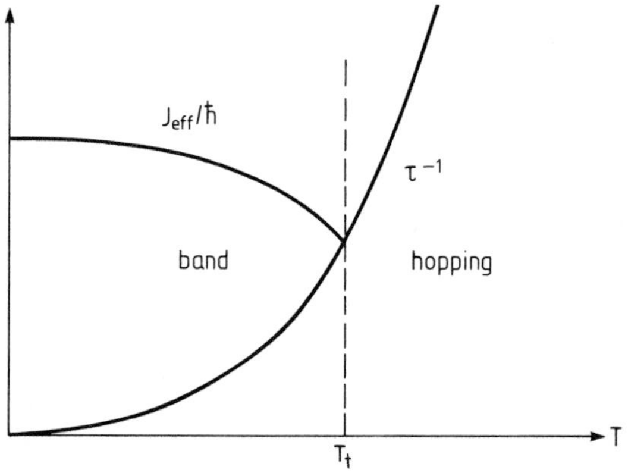

Fig.2.5. Qualitative behavior of the coherent transfer rate given by $J_{eff}/\hbar$ and the decay rate of the band state τ^{-1} as a function of temperature

Maintaining the criterion that the transition takes place at the temperature where the incoherent rate exceeds the coherent one, further scattering mechanisms like the intrawell scattering of KAGAN and KLINGER /2.29/ reduce T_t. For this case, KEHR /2.33/ has given an explicit expression for the transition temperature

$$T_t \cong \frac{\hbar c}{k}\left(\frac{648\,\pi^6 \rho^4 c^{10} \hbar^4}{7!8!\,\Omega^2 d^6 \Delta C^2 P^4}\right)^{1/16} \qquad (2.30)$$

which does not depend explicitly on J_{eff}. For Cu a value of $T_t = 34$ K was estimated, which is in the order of a tenth of the Debye temperature. In this respect Kehr's result is also characteristic of other calculations of T_t in the small-polaron framework, which consider only phonon scattering and typically yield transition temperatures of some tenth of θ_D /2.37,38/. A further reduction of T_t could be expected if other scattering mechanisms like electron or impurity scattering become important.

Finally we note that the independence of T_t on J_{eff} implies that depending on the magnitude of J_{eff} coherent transfers may be prevalent even for coherency lengths $\ell = v\tau = J_{eff}d\tau/\hbar$ much smaller than typical lattice distances. This was first stated by KAGAN and KLINGER /2.29/, who argued that in the case of strong coupling or narrow bands even for scattering rates much larger than the bandwidth coherent processes could still play the dominant role. While HOLSTEIN's /2.2/ criterion for T_t, namely $J_{eff} \cong \hbar/\tau$ or $J_{eff}\ \tau/\hbar \cong 1$, is in agreement with the usual picture of band-like transport which breaks down when the coherency length ℓ becomes smaller than a typical lattice distance, the small-polaron results in the spirit of KAGAN and KLINGER /2.29/ extend the range of "band transport" to still higher scattering rates.

2.6 Capture and Release from Trapping Centers

Impurities and defects create (i) long-range, but weak disturbances which may influence the elementary step of small-polaron motion (Sects.2.4,5.1) and (ii) locally they may distort the host lattice heavily, providing interstitial sites with strongly lowered ground-state energies. These sites are capable of trapping the diffusing interstitial and of keeping it in "custody" until an escape by thermal activation is possible. In general one impurity creates a number of such trapping sites. We call the whole configuration of trapping sites created by one lattice defect a structured trap.

The strength of the binding at a trap may origin from elastic interaction but could also be determined by direct "chemical" forces. For our perspective the details of the actual trapping mechanism are not of prime interest, but the diffusion process under the influence of trapping centers will be addressed.

We will describe transport in the presence of traps in a two-state model /2.39, 40/. In this model a particle spends an average time τ_1 in a mobile state where it can diffuse through the lattice. Thereafter it is captured by a trap, and released from it after an average time τ_0. A generalization of such a model to trapping on more than one kind of traps is obvious. Such an extension, however, does not change its general features and it suffices to discuss τ_1 and τ_0. The inverse lifetimes $1/\tau_1$ and $1/\tau_0$ are the capture and escape rates associated with the trapping process. The consequences of trapping processes for the experimental observation of diffusion will be discussed in the respective experimental Sects.3.6 and 4.2.3. Here we consider in some detail the parameters τ_1 and τ_0.

Since the ground-state energy of a particle in a trap is lowered with respect to the undisturbed lattice, escape is possible only by thermal activation, and the temperature dependence of τ_0^{-1} has to exhibit an activated behavior. The activation energy should be specific for each trapping center, i.e., characteristic for each kind of defect or impurity present. Since it is a local property of the trap, τ_0^{-1} should not depend on the concentration of the impurities. Its prefactor is not simply related to the jump rates in the undisturbed regions of the lattice – the overlap integral can be expected to be sensitive to changes in the potential surfaces and will be in general different from its undisturbed value.

For structured traps, where the escape process occurs over several steps, no simple relation between the apparent activation energy, measured in a neutron or μSR experiment, the binding energy at the trap and the energies involved in the different steps can be anticipated. A similar statement holds for the prefactors /2.4/.

The capture rate τ_1^{-1} is determined by the nature of the transport processes which lead the particle into the trap. In the case of phonon-assisted diffusion where a localized particle performs jump diffusion, the capture process is diffusion limited. In a continuum approach for infinitely strong sinks τ_1^{-1} has been evaluated to /2.42/.

$$\tau_1^{-1} = 4\pi r_t D n_t \quad , \tag{2.31}$$

where r_t is the trapping radius. For a random walk on a lattice with randomly distributed traps and allowing for escape processes, a similar result has been derived /2.40/. For structured traps r_t is an effective radius, which for classical diffusion measures the distance from the impurity to the interstitial position, where the saddle-point energies are lowered by approximately kT compared to the undisturbed lattice. Taking into account the $1/r^3$ dependence of the elastic interaction $r_t \sim T^{-1/3}$ results /2.43/. For quantum diffusion the saddle-point loses its meaning and a criterion for r_t has to consider changes in the coincidence energies or energy differences between sites, caused by the interaction with the trapping center, but this criterion has not yet been worked out in detail.

As has been pointed out by HODGES /2.44/ for bandlike transport with coherency lengths $\ell \gg r_t$, the capture process is capture limited and the capture rate is given by

$$\tau_1^{-1} = \sigma v n_t \quad , \tag{2.32}$$

where σ is the absorption cross section of the trap. For long mean free paths and large muon wavelengths, which may be achieved at sufficiently low temperatures where the de Broglie wavelength of the muon is large, the cross section becomes proportional to $1/v$ in the case of deep traps. Then the temperature dependence cancels, and τ_1^{-1} becomes solely a function of the impurity concentration, as has been observed, e.g., for positron trapping at vacancies /2.45/.

Though the positron has a much lighter mass than the muon, muon experiments are performed at much lower temperatures so that similar conditions with respect to the thermal wavelength may hold. For shallow traps or intermediate regions with $r_t \cong \ell$, no simple temperature dependence can be predicted. However, a smooth crossover between the two regimes can be expected. For the proton, band transport is not likely and we expect to be always in the diffusion-controlled regime.

2.7 Small-Polaron Formation and the Possibility of a Delay to Self-Trapping

So far we have always assumed that the muon or the proton and its isotopes are present in a self-trapped small-polaron state. For H and its heavier isotopes there is no doubt about the existence of local lattice deformations which characterize the self-trapped state, while the positron is known not to form polaronic states in metals. Though for the muon with its mass much nearer to the proton than to the positron, the ground state should still be self-trapped, questions have been raised /2.46,47/ whether delays to self-trapping may influence muon behavior during the observation period which is limited by its finite lifetime. Therefore, at the end of the theoretical section on small-polaron transport, we shall briefly discuss the theoretical approaches to the self-trapping process.

As has been shown at the beginning of Sect.2.1, an interstitial particle gains potential energy by host lattice relaxation. However, localization into a small-polaron state increases its kinetic energy, since Fourier components corresponding to higher momentums are involved in the wave function of a localized state compared to a long wavelength plane wave.

For the case of the positron the balance between kinetic energy and potential energy has been investigated quantitatively in the framework of continuum theory /2.48/ as well as of harmonic lattice theory /2.49/. Both approaches lead to a kinetic energy contribution up to one order of magnitude too large in order to allow self-trapping. On the other hand, since the kinetic energy varies as $1/M$ while the elastic energy depends only weakly on M for the muon as for the proton, the ground state should be the small-polaron state.

The next problem we have to address is the question how the muon, which is implanted and initially starts in a non-self-trapped, delocalized state, reaches its equilibrium state. For the adiabatic limit, where the light particle always immediately adjusts itself to the strain pattern of the lattice, EMIN /2.50,51/ investigated the existence of barrier to self-trapping using scaling arguments. Following him, one can write the energy of a particle added to an elastic continuum as a function of a scaled position variable with scaling factor R describing the size of the particle wave function.

Three components contribute in scaling: (i) the kinetic energy which scales as T_e/R^2, (ii) the elastic potential energy of the particle which behaves as $-V_{int}/R^3$, and finally (iii) the resulting strain energy of the deformed continuum which is given by $V_{int}/2R^3$. Thus, the total energy of the particle which scales as

$$E(R) = T_e/R^2 - V_{int}/2R^3 \quad , \tag{2.33}$$

always exhibits a maximum for finite R (Fig.2.6). Both an infinitely spread out muon as well as the small polaron, which in a discrete system shrinks to typical interatomic distances, appear to be possible and are separated by a barrier. Similar results have also been obtained by BROWNE and STONEHAM /2.46/ using slightly different argumentation. In the context of scaling it can easily be seen that impurities which interact elastically with the muon (interaction potential $-V_{int}^{imp}/R^3$) reduce the barrier height.

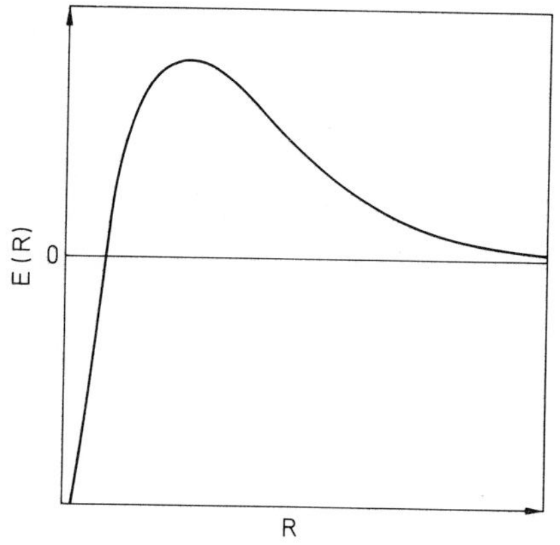

Fig.2.6. The adiabatic ground state energy of an elastic continuum and an added light interstitial is plotted against the scaling factor R /2.47/

To see whether such an extended state exists for any significant amount of time, the dynamic stability of this non-self-trapped state has to be investigated. Qualitatively, it is clear that the ratio A of the muon velocity in the free state characterized by the corresponding tunneling matrix element J_f and of the reaction time of the lattice measured by the vibrational frequencies ω_D of the host lattice will be the important parameter which determines the stability of the extended state

$$A = zJ_f/\hbar\omega_D \quad . \tag{2.34}$$

For small A the lattice can react and the muon will be self-trapped within a vibrational period $\sim 10^{-12}$ s, while for large A self-trapping will become increasingly difficult. This simple qualitative picture is supported by detailed variational studies of an electron in a deformable medium beyond the adiabatic limit /2.50/. As an additional criterion these calculations reveal that for a given A free and polaron states can coexist only within a certain range of $|\Delta U|/\hbar\omega_D$.

Now to judge whether such metastable extended states may play any role in muon dynamics we follow a suggestion of EMIN /2.52/, whereby in a first approximation the rigid lattice bandwidth of the muon can be estimated by treating it like a free electron in a periodic lattice. For Al, e.g., a bandwidth of about 30 meV results, which is of the same magnitude as the typical lattice frequencies. Thus A is about one, and we are confronted with a borderline situation where the theory cannot predict whether self-trapping will occur immediately or whether intermediate metastable states are possible.

With respect to the actual experiments, the influence of temperature on the self-trapping mechanism is of great interest. Here it is useful to follow the picture given by BROWNE and STONEHAM /2.46/ of a critical fluctuation q_c of the host atoms above which localization occurs. For a single oscillator the probability that its vibrational coordinate exceeds a certain value q_c is given by [for the asymptotic form cf. (2.13)]

$$P_T(q > q_c) = \mathrm{erfc}(q_c/\tilde{q}) \quad , \tag{2.35}$$

where $\tilde{q} = \sqrt{(\hbar/2M\omega)\coth(\hbar\omega/2kT)}$. Expressing q_c by the bandwidth zJ_f and the coupling energy ΔU one finds

$$P_T(q > q_c) = \mathrm{erfc}\left(\frac{zJ_f}{\sqrt{\Delta U\ \hbar\omega\coth(\hbar\omega/kT)}}\right) \quad . \tag{2.36}$$

Equation (2.36) shows again that a barrier to self-trapping is significant only for bandwidths of the order of, or larger than, typical lattice frequencies ω_D. In addition, temperature influence can be expected only for thermal energies kT of the order of $\hbar\omega_D$. This basic result does not change if one considers the strains arising from the distribution of thermal phonons in a crystal, e.g., for Al BROWNE and STONEHAM /2.46/ calculated a 14% increase of the thermal strain energy between T = 0 and room temperature. Hence, thermal fluctuations are unlikely to provide temperature-dependent formation mechanisms at low temperatures.

So far all considerations on the existence of metastable states which could delay the self-trapping process fall short in providing calculations of actual transition rates, but consider more or less the static situation "barrier to self-trapping", or give criteria for, e.g., the magnitude of A or the temperature or impurity concentration which could delay small-polaron formation. An explicit treatment of the self-trapping kinetics is still missing.

3. Muon Diffusion Experiments in Metals

3.1 The Muon as a Light-Hydrogen Probe

Analysis of cosmic ray cloud chamber patterns in 1936 and 1937 led to the conclusion that particles less massive than protons but more penetrating than electrons exist /3.1.2/. The muons were discovered. Early assumptions related it to the nuclear forces proposed by YUKAWA /3.3/. Later on, however, it was learnt that the pion and not the muon is the primary particle in cosmic radiation and that the pion is related to Yukawa's nuclear forces. Nowadays we know that the muons, which can be positive or negative, fall into the class of leptons, together with electrons, neutrinos and the newly discovered $\tau^{\pm}$ particles.

After a mean lifetime of $\tau_\mu = 2.2\ \mu s$ the muon decays into a positron and two neutrinos

$$\mu^+ \rightarrow e^+ + \bar{\nu}_\mu + \nu_e \quad . \tag{3.1}$$

In 1957, twenty years after the discovery of the muon, GARWIN et al. /3.4/ observed the violation of parity in the decay of the muon which manifests itself in a correlation between the directions of e^+ emission and of the muon spin in the moment of decay. This anisotropy of the decay positrons is the basis of muon applications in solid-state physics. Studying positron distributions in a time-differential fashion allows direct observation of muon spin dynamics under the influence of the various magnetic fields characteristic for the piece of matter where the muon has been stopped. The muon becomes a microscopic probe for the investigation of solid-state phenomena.

On an ensemble of polarized muons implanted in a metal essentially two quantities can be measured: the muon spin precession frequency and/or the decay of muon polarization. They are related to static as well as dynamic features of the solid under investigation: in magnetic materials the precession frequency reveals information on the local field at the interstitial muon site similarly to Mößbauer and NMR experiments which yield the local fields on substitutional sites. In nonmagnetic materials Knight shift measurements cast light on the local electronic structure around the muon. The investigation of the field-dependent decoupling of electronic and magnetic interactions allows the determination of the electric field gradient (EFG) created by the muon on its neighboring atoms, as well as the assignment of the muon location /3.5,6/.

The time scale, which is probed by the muon, is related to the magnetic field fluctuations $\langle \Delta B^2 \rangle$ in the solid and the lifetime of the muon. In nonmagnetic substances, where $\langle \Delta B^2 \rangle$ is caused by nuclear dipole fields, the resulting time scale amounts to $10^{-7} < t < 10^{-5}$ s, in magnetic materials it can be extended down to 10^{-12} s. Time-dependent field fluctuations may occur either due to dynamic processes in the host material like spin relaxation in spin glasses or by the motion of the muon itself, which is the primary concern of this review.

Though the mass of the muon is lighter by a factor of 9 than the proton, it is still heavier by a factor of 207 than the electron or positron. Furthermore the electrostatic interaction between host and proton or positive muon can be considered to be very similar. Thus, with respect to transport mechanisms of light interstitials in metals, the positive muon provides us with a light isotope of hydrogen which allows us to investigate the isotope effects in hydrogen diffusion toward lighter masses, where quantum effects are expected to be more pronounced. In addition, because muons are implanted, temperature ranges and materials are accessible, where H diffusion studies are impossible due to low solubility. Finally, muon experiments are undertaken at infinite dilution. Therefore, interaction effects between the light interstitials which may inhibit quantum diffusion for the case of H /2.29/ do not play any role for the muon.

3.2 Basic Muon Properties

3.2.1 Muon Production

Positive muons are created by the weak interaction decay of positive pions

$$\pi^+ \rightarrow \mu^+ + \nu_\mu \tag{3.2}$$

where the lifetime of the charged pion amounts to $\tau_\pi = 2.6 \cdot 10^{-8}$ s. The intrinsic properties of the muon are summarized in Table 3.1 and compared with those of the

Table 3.1. Intrinsic properties of muon and proton

Particle	Mass	Spin	Magn. moment	Lifetime
Muon	105.659 MeV/c^2	1/2	3.18334 μ_p	τ_μ = 2.197 μs
Proton	938.28 MeV/c^2	1/2	μ_p = 2.7928 μ_N	stable

proton. In the rest frame of the π^+ the muons have a fixed energy of 4.1 MeV corresponding to a linear momentum of p_μ = 29.8 MeV/c. Since the π^+ possesses no spin and the neutrino ν_μ has a negative helicity, the conservation of angular momentum also requires a longitudinal polarized muon with spin direction opposite to its linear momentum. In the standard experimental setup, where pions decay in flight in so-called muon channels, polarization in the laboratory frame is reduced to about 75% for kinematic reasons.[2] Typical μ^+ momenta are between 100 and 300 MeV/c. In order to achieve maximum stopping of muons in the sample they are degraded in energy before the target. High stopping rates are then obtained by samples offering a few grams per cm^2 beam cross section. Figure 3.1 shows the experimental realization at the muon facility at CERN which is centered around the 600 MeV synchrocyclotron (SC). The main beam is obtained from π^+ production in an external target and collected in the C beam line. At the normal momentum selection it has a maximum intensity of $3 \cdot 10^4$ μ^+/s over an area of 8 cm^2 and a polarization of 75% - 80%. A second μ^+ beam uses an internal target for π^+ production and is taken out in the beam line LJ1. It has only about half the intensity of the main beam but its e^+ contamination is much smaller which makes it particularly useful for forward-backward geometry experiments.

An alternative possibility of μ^+ production constitutes the so-called surface or Arizona beam, where the pions decay at rest at the surface of the production target. In this case the beam polarization is 100% and the linear momentum of the μ^+ amounts to 29.8 MeV/c originating from the π^+ decay itself. At these low energies a sample thickness of typically 0.1 mm is sufficient. The advantage of small samples has to be traded for difficulties with, e.g., windows in the beam and strong aberrations in applied magnetic fields.

[2] The muons are extracted out of the pion beam under a certain angle. Taking into account the large momentum of the π^+ compared to that of the muon in the rest frame, it is obvious that muons out of an angular range in the rest frame can contribute to the extraction direction, resulting in reduced polarization.

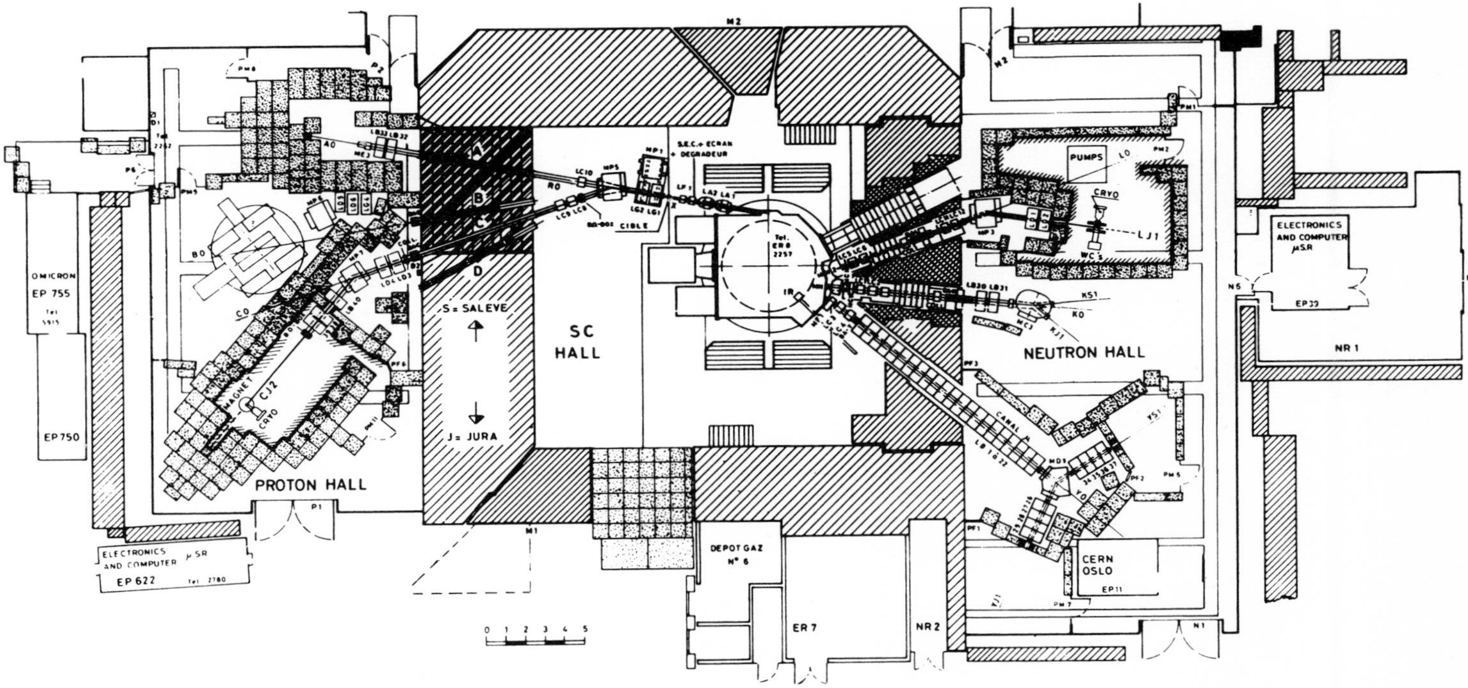

Fig.3.1. Beam lay-out for the two μSR experimental stations at the 600 MeV synchrocyclotron at CERN

3.2.2 Asymmetry and Decay

Muon decay (3.1) is governed by the weak interaction and does not conserve parity. The resulting positron spectrum can be derived rigorously from the 4-fermion current-current interaction /3.7/ and is given by

$$\frac{dN(\omega,\vartheta)}{d\omega d\Omega} = \frac{\omega^2}{2\pi}\left[(3-2\omega) - P(1-2\omega)\cos\vartheta\right] \tag{3.3}$$

$$= \frac{c}{2\pi}(1+A\cos\vartheta) \quad , \tag{3.4}$$

where $N(\omega,\vartheta)$ is the number of positrons of the energy $\omega = E/E_{max}$ emitted under an angle ϑ with respect to the direction of the muon spin. Here P is the muon polarization and $E_{max} = M_\mu/2 = 52.8$ MeV. Due to the high average positron energy of $\bar{E}$ = 36 MeV the positrons can easily escape the sample and are detected with no difficulty. The asymmetry parameter is $A = P(2\omega-1)/(3-2\omega)$ and varies between -P/3 and P. The positrons are counted without energy discrimination but the efficiency of detection $\varepsilon(\omega)$ depends on energy

$$\frac{dN}{d\Omega} = \int_0^1 \frac{dN(\omega,\vartheta)}{d\Omega d\omega}\,\varepsilon(\omega)d\omega \quad . \tag{3.5}$$

For $\varepsilon(\omega) = 1$, $\bar{A} = P/3$ would follow. Since low-energy positrons, where $A(\omega)$ is negative, are detected with lower efficiency, the actual asymmetry is generally larger than P/3. This effect, however, is counterbalanced by finite detection angles and the experimentally achieved asymmetries $A_{exp} = aP$ typically reaching values of $a \cong 0.3$. Figure 3.2 illustrates the asymmetric positron distribution in comparison with the isotropic case.

3.3 *Experimental Setup*

The basic experimental setup for a muon spin rotation (μSR) experiment is as old as the detection of the parity violation in μ^+ decay /3.4/, where the oscillation of the positron emission probability due to the Larmor precession of muons stopped in carbon has been studied. Since then, the aim of a μSR experiment has remained to determine the time evolution of the polarization of muons stopped in matter. For this purpose knowledge about the time which the muon spends in the sample is required. Most commonly this is achieved by measuring the individual lifetime of each muon: the incoming muon starts a clock and the decay positron stops the clock when it hits a detector telescope at a certain angle with respect to the beam direction.

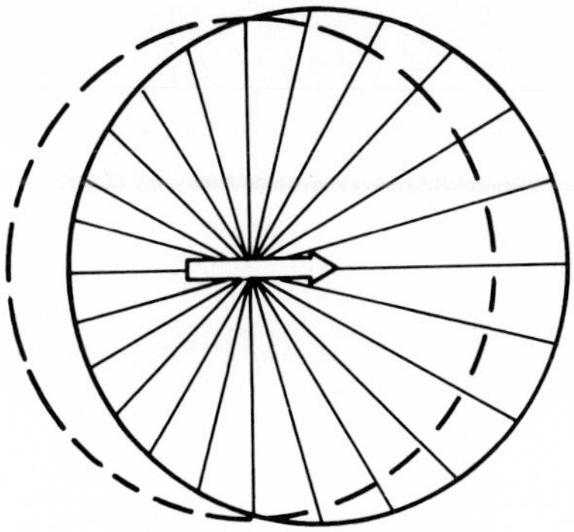

Fig.3.2. Plot of the asymmetric positron distribution $(dN/d\Omega)(1+A\cos\vartheta)$ with $A = 0.3$

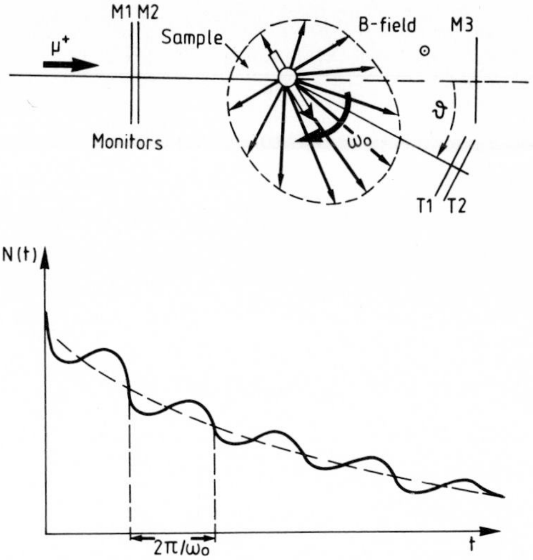

Fig.3.3. (Top) Schematic sketch of a transverse μSR setup. (Bottom) Time dependence of the resulting e^+ count rate

This approach allows only one muon at a time in the sample. Alternatively, one could think about the utilization of μ^+ bursts with a halfwidth short compared to τ_μ. Such a mode of operation is actually used at KEK in Japan where the corresponding proton accelerator has the necessary time structure. Another more specialized approach constitutes the so-called stroboscopic method[3] which is mainly applied for precession frequency measurements.

During an experiment typical 10^6-10^7 events are recorded and arranged in a time histogram. The number of positrons counted in a certain time interval measures the direction and magnitude of the muon spin polarization at the time these muons decayed, since the e^+ emission probability is peaked in the μ^+ spin direction. Figure 3.3 shows schematically a μSR setup with transverse field geometry, which is by far the most common one for metal studies. The applied magnetic field is perpendicular to the plane opened up by the beam and the telescopes. Two monitors in front of the sample count in coincidence the incoming muons. If no signal has been obtained from the counter M3, at the same time the muon has stopped in the sample and the electronic clock is started. The emitted positron is identified by coincidence signals coming from T1 and T2, and stops the clock. The incoming muon, which retains its polarization during the fast thermalization process in a metal /3.9/, starts to pre-

[3] The stroboscopic method applies a muon beam with periodically modulated intensity in synchronization with the Larmor frequency of the implanted muons. Muons of different pulses add with correct phases and precess, independent of their arrival time, more or less coherently /3.8/.

cess with a Larmor frequency ($\omega_\mu = 2\pi\gamma_\mu B$, γ_μ = 13.55 KHz/Gauss) and the resulting time histogram is shown schematically in Fig.3.3. The muon decay function is superimposed by an oscillatory component which has a maximum whenever the μ^+ spin points in the direction of the counter. For the time-dependent counting rate we have

$$N(t) = N_0 \exp(-t/\tau_\mu)\left[1+aP(t)\cos(\omega_\mu t+\phi)\right] \quad , \tag{3.6}$$

where a is the experimentally achieved asymmetry and ϕ is a phase angle depending on the geometry of the experiment. It allows the determination of the precession frequency as well as the damping of the spin polarization P(t). The range of internal fields accessible to the experiment lies between 20 G (2 precessions in a time window of 10 µs) and 50 kG limited by the time resolution of the electronics. The accuracy can be pushed to a few ppm. In transverse geometry, the damping of the muon polarization P(t) is caused by random fluctuations of the local field originating, e.g., from the nuclear dipole moments in a nonmagnetic material. Muons stoped at different sites experience different Larmor frequencies and lose their phase coherence. No energy transfers are involved in this depolarization process.

The damping of the muon polarization corresponds to a linewidth and a characteristic line shape in the frequency spectrum (after Fourier transformation). The whole phenomenology of µSR analysis is identical to that of NMR. For an immobile muon the line shape is Gaussian

$$P(t) = P_0 \exp(-\sigma^2 t^2) \quad , \tag{3.7}$$

where σ^2 is related to the second moment of the frequency distribution due to internal fields. For a mobile muon the field fluctuations are averaged and motional narrowing occurs. The depolarization then has the form

$$P(t) = \exp(-2\sigma^2 \tau_c t) \quad , \tag{3.8}$$

where τ_c is the average time during which correlations between the frequencies exist. The time $T_2 = 1/2\sigma^2\tau_c$ is also called spin-spin relaxation time. Further, τ_c is proportional to the mean residence time τ of the muons at their interstitial sites. For the whole range of mobilities a formula, introduced originally to evaluate line-shape profiles of diffusing particles in NMR /3.10/, can be applied (Sect.3.5.1):

$$P(t) = \exp\{-2\sigma^2\tau_c^2\left[\exp(-t/\tau_c) - 1+t/\tau_c\right]\} \quad . \tag{3.9}$$

Longitudinal and zero field experiments are performed in forward-backward geometry with the magnetic field in or reverse to the beam direction. The emitted positrons are counted under 0° and 180° with respect to the beam, and using (3.6) the longitudinal depolarization function $P_\ell(t)$ is simply given by

$$aP_\ell(t) = \frac{N_f - \alpha N_b}{N_f + \alpha N_b} \quad , \tag{3.10}$$

where α accounts for possible asymmetry of the instrument. Other than transverse depolarization longitudinal decay involves reversal of the muon spin in the applied external field B_0 which is associated with a transfer of the Zeeman energy $\Delta E = 2\mu_\mu B_0$ to the lattice. We speak of spin lattice relaxation, and the polarization decay has the form

$$P_\ell(t) = e^{-t/T_1} \quad , \tag{3.11}$$

where T_1 is the so-called spin lattice relaxation time. For longitudinal relaxation in zero and low-field NMR a stochastic theory has been worked out by KUBO and TOYABE /3.11/ and applied to µSR by YAMAZAKI /3.12/. Under these circumstances the μ^+ spins evolute under the influence of the randomly distributed dipolar fields, and in longitudinal direction the transverse depolarization function is also observed. In particular, this approach is more sensitive to long relaxation times and allows a decoupling of spin-spin and spin-lattice relaxation which is not possible under transverse geometry. Another interesting feature of the zero field method has been pointed out by PETZINGER /3.13/. For an immobile muon after an initial decay $P_\ell(t)$ increases asymptotically for long times to 1/3 of its initial value, since precession due to longitudinal components of B_{dip} does not change P_ℓ on the average. For a muon undergoing trapping processes this should enable one to distinguish whether a particle diffuses toward a trap or escapes from it. Only when the muon ends up in a trap should the restoration of $P_\ell(t)$ be observed.

3.4 Static Linewidth – Muon Location and Electric Field Gradients

A precondition for a quantitative interpretation of µSR results, e.g., on Knight shifts, internal fields and diffusion properties is knowing the muon lattice location. Furthermore, since in dilute systems the H isotopes generally occupy identical sites, it appears to be a reasonable assumption that H and μ^+ also occupy the same lattice sites and thus site determinations for the muon might help to complement the still rather limited data on H location.

The clues to solving the problem are the field and orientation dependence of the static dipolar width. In a metal host which possesses sufficiently large nuclear dipole moments like the fcc metals Cu and Al or the bcc metals Nb, Ta or V, the dipole fields at the muon sites are in the order of a few Gauss. Above 1 mK they are oriented randomly and cause a magnetic field distribution around the value of the applied field B_0. It is characterized by its second moment $\langle \Delta B^2 \rangle$ and gives rise to the static linewidth σ^2 in the transverse depolarization function (3.7). In analogy to NMR for nuclear dipoles the second moment is given by the VAN VLECK formula for nonidentical spins /3.14/

$$\sigma^2 = \frac{\hbar^2}{6} \gamma_\mu^2 \gamma_I^2 \, I(I+1) \sum_j \frac{(1-3\cos^2\vartheta_j)^2}{r_j^6} \quad , \tag{3.12}$$

where r_j is the vector from the muon to the nuclear spin I_j and ϑ_j is the angle between $\underline{r}_j$ and the external field B_0, γ_I and γ_μ are the gyromagnetic factors for the nuclei and the muon respectively. In a single crystal σ^2 depends on the crystal orientation relative to B_0. Table 3.2 presents theoretical σ^2 values for octahedral and tetrahedral sites in fcc and bcc structures taking Al and Nb as examples. For bcc metals magnetically inequivalent tetrahedral or octahedral sites are present. Therefore, more than one Gaussian line appears in certain directions. These so-called Van Vleck values are distinctly different for different crystal orientations and characteristic for a particular interstitial site. Their measurement should be sufficient to determine the interstitial site.

However, first experiments on Cu, applying low transverse fields, showed very little differences of the σ values for the different crystal directions /3.15/. Subsequently HARTMANN pointed out that high transverse fields have to be applied in order to reach the σ values given by the dipole sums /3.16/. The muon as a charged impurity creates to first order a radially directed electric field gra-

Table 3.2. Theoretical static linewidths σ [μs^{-1}] for muons in Al and Nb

Host metal		Field direction		
		100	110	111
Al (fcc)	octahedral	0.31	0.16	0.07
	tetrahedral	0.08	0.28	0.32
Nb (bcc)	octahedral	0.740 (1) 0.380 (2)	0.205 (2) 0.459 (1)	0.197 (3)
	tetrahedral	0.188 (1) 0.449	0.124 (1) 0.393 (2)	0.308 (3)

dient (EFG) at the nearby nuclei. If these nuclei posses a quadrupole moment, then the direction of the external field is not longer the unique quantization axis for the nuclear moments, and the orientational dependence of the dipolar width is changed drastically. The correct expressions for σ have to be calculated with the nuclear spins I undergoing combined electric and magnetic interactions /3.17/ (App.3). The result depends strongly on the ratio of magnetic and electric interaction b. For small b the electric interaction dominates, and σ is nearly isotropic. For $b \gg 1$ the Van Vleck values are recovered. The crossover from electric to magnetic behavior occurs around b = 5. Its experimental observation allows determination of the EFG at the surrounding nuclei.

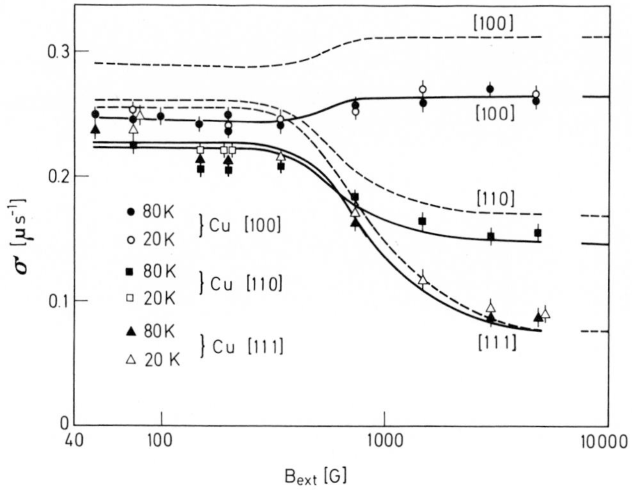

Fig.3.4. Field and orientational behavior of σ observed in Cu. HARTMANN's theory /3.16/ (octahedral sites) for a rigid lattice (-----) and allowing for lattice relaxation (———)

The first experiment on the field and orientation dependence of the static linewidth has been undertaken on Cu /3.18/. Figure 3.4 compares the result with HARTMANN's theory /3.16/. It is obvious that the muon similar to H in other fcc metals like Pd occupies an octahedral site in Cu. Thereby the absolute value of the static linewidth is reduced relative to an unrelaxed lattice. The observed σ values correspond to a 5% displacement of the nearest Cu neighbors. The crossover from electric to magnetic behavior determines the quadrupole frequency of the nearest neighbors to $\omega_E/2\pi$ = 0.18 MHz corresponding to an electrical field gradient at the nearest-neighbor Cu atoms of $q = 0.30 \pm 0.15$ Å^{-3}. The large error stems from the 50% uncertainty in the Cu quadrupole moment.

The second well-investigated metal is Al, where the first measurement of field dependence was carried out on a single crystal containing 1300 ppm Mn /3.19/. While

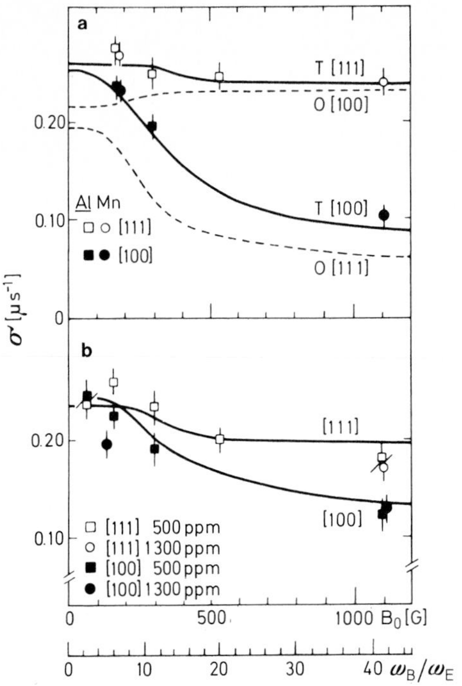

Fig.3.5

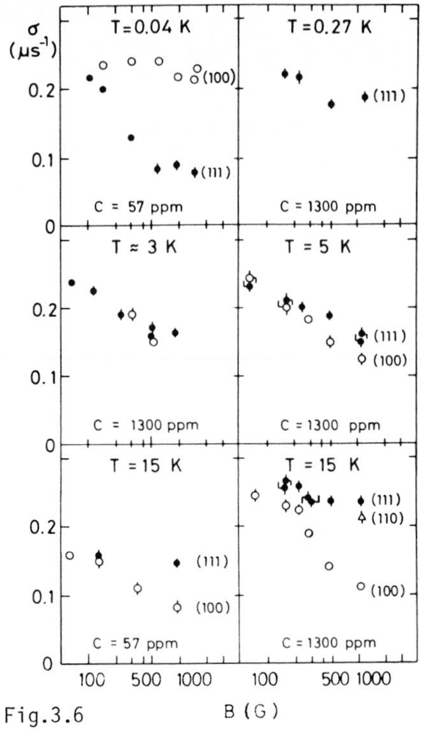

Fig.3.6

Fig.3.5a,b. Field dependence of σ for $AlMn_{1300\ ppm}$. (a) T = 15 K, (——) tetrahedral sites; (-----) octahedral sites. (b) T = 5 K, theoretical curves are a sum of 30% octahedral and 70% tetrahedral occupation

Fig.3.6. Field dependence of σ for various AlMn samples and various temperatures. The 0.04 K data show an octahedral site symmetry, while the two measurements at 15 K show a tetrahedral site occupancy

at 15 K the muon site is tetrahedral (Fig.3.5a), already the first measurements revealed the puzzling result that below 15 K the data follow neither the curve for tetrahedral nor that for octahedral site occupancy, but seem to be a mixture of both (Fig.3.5b). Since then several other measurements on single crystals have been carried out /2.24/ and the present situation is illustrated in Fig.3.6. These data reveal an octahedral site occupancy at 40 mK, while at intermediate temperatures a mixture occurs, and finally at 15 K a tetrahedral symmetry is found as before. This change of site at very low temperatures indicates a delicate balance between the energies associated with the octahedral and tetrahedral sites, and finds its counterpart in the observation of a tetrahedral site occupancy for H in Al which seems to be associated with a vacancy /3.20/. We shall address this feature again in Sect. 3.6 in the context of diffusion and trapping.

As in the case of copper, the observed σ values are smaller than the calculated values for muons in an undistorted lattice. For $AlMn_{1300\ ppm}$ this effect amounts to about 23%, corresponding to a 7% local lattice expansion. The quadrupole frequencies for tetrahedral and octahedral sites were determined to 0.051 MHz and 0.038 MHz respectively. The corresponding electric field gradients are q = 0.4 $Å^{-3}$ and q = 0.3 $Å^{-3}$ with similar uncertainties as in copper. The smaller value at the octahedral site may just reflect the increase in the Al μ^+ distance which, assuming a $1/r^3$ dependence of the EFG would bring about a reduction to 65% of the tetrahedral value.

The previous discussion implied a muon in a pure aluminum environment not trapped very closely to a Mn impurity. This assumption is based on the fact that Mn in Al itself exhibits a strong EFG on neighboring atoms (q = 0.45 $Å^{-3}$) /3.21/. Therefore a muon, being a next neighbor to a Mn atom, would experience a combined electric and magnetic interaction of a symmetry very different to that of a radially directed EFG originating from the muon. However, the exact behavior of the muon linewidth in this complicated situation with two different sources of field gradient has not yet been calculated.

Though the experimental results for the EFG are affected by considerable uncertainties due to the limited accuracy of our knowledge about the nuclear quadrupole moments, they are of theoretical interest since they represent EFG values very close to the charged impurity and allow critical tests of electronic screening calculations. Recent calculations by JENA et al. /3.22/ using the self-consistent formalism of HOHENBERG et al. /3.23/ arrived for Cu at q = 0.26 $Å^{-3}$, in good agreement with the experiment. Their prediction for Al, however, of q = 0.13 $Å^{-3}$ for an octahedral site is out by more than a factor of two.

Finally the observed local lattice expansions allow conclusions on the strength of the elastic muon lattice interaction which is one of the basic parameters of the small-polaron theory.

Describing the dynamic properties of the lattice by one force constant $f = C_{44}a$ /3.24/, where C_{44} is an elastic force constant and a is the lattice parameter of the host, the magnitude of the Kanzaki force ψ has been estimated from the requirement of an energy balance between the work done by ψ and the elastic energy stored in the springs connecting the host atoms. For octahedral and tetrahedral sites in an fcc structure one gets /3.25/

$$\psi = 4\ f\delta R \quad . \tag{3.13}$$

For Cu a value of $\psi = 1.05 \cdot 10^{-4}$ dyn has been evaluated. This yields a double force tensor $P = \psi a = 2.36$ eV in reasonable agreement with TEICHLER's result obtained from a nonlinear screening approach of $\psi a = 2.0$ eV /2.17/ and $\psi a = 2.65$ eV estimated from an analysis of the μ^+ diffusion data in terms of small-polaron theory /2.16/. Using the same approach for Al for a tetrahedral site occupancy the following results can be extracted: $\psi = 0.87 \cdot 10^{-4}$ dyn, $P = \psi a/\sqrt{3} = 1.26$ eV; and $\Delta V/\Omega = 20\%$.

Concerning the bcc metals Ta, Nb and V, the experimental situation is by far less satisfactory: (i) because of the magnetic inequivalence of the interstitial sites of one category, the data analysis is more complicated than in fcc metals and has up to now not been conducted properly, (ii) the large quadrupole moment of Ta provides a very strong electric interaction which cannot be quenched by external magnetic fields and consequently the Van Vleck values cannot be reached, (iii) in all three metals the absolute value of σ depends strongly on the impurity concentration present in the host. However, while in Ta and Nb plateaus of σ over certain temperature intervals exist, such a saturation has not been observed for V. In view of these difficulties information on muon location in bcc metals is still limited, though there seems to be a general tendency toward tetrahedral site occupancy which is in agreement with the results on H in these materials. In vanadium FIORY et al. /3.26/ tried to assign a tetrahedral site for the muon using the absolute value of σ in the polycrystalline sample, where the orientationally averaged σ is only about 7.5% different for both sites. Considering the possible effect of lattice expansion and impurities /3.27/, this approach is unsatisfactory. Recently SCHILLING et al. /3.28/ reported field-dependent measurements on a V single crystal which follow qualitatively the behavior expected from a tetrahedral site. The authors fitted the data quantitatively assuming a mixed occupancy of 81% tetrahedral and 19% octahedral sites. The absolute values of σ are nearly a factor of 2 below the theoretical expectation and are presumably a consequence of only partially localized muons. For the electric field gradient a value of $q = 1.1$ Å^{-3} was extracted.

In niobium already the first field-dependent measurements by LANKFORD et al. /3.29/ and HARTMANN et al. /3.30/ gave strong hints for a tetrahedral site occupancy. For the high field ratio of $\sigma^{100}/\sigma^{110}$ the experimental results were 1.2 /3.29/ and 1.35 /3.30/ which have to be compared with the theoretical expectations of 1.24 for a tetrahedral and 2.79 for an octahedral site. Later on, measurements of BIRNBAUM et al. /3.31/ in (100) and (110) directions were interpreted quantitatively by SCHILLING et al. /3.28/, again using mixed tetrahedral and octahedral site occupancies (67% : 33%), the EFG came out to $q = 1.4$ Å^{-3}.

Finally, for Ta a field or orientation dependence of σ could not be detected /3.32/. This is in agreement with an overwhelming electric interaction, which causes a nearly isotropic static width and prohibits a site determination. The absolute value of σ alone does not allow the assignment as discussed above.

3.5 Motional Narrowing, Muon Diffusion

3.5.1 Theory of Depolarization

In a transverse μSR experiment jump diffusion of localized muons manifests itself in a reduced damping of the muon precession – motional narrowing occurs. Here we consider a nonmagnetic solid, where the internal fields are generated by the nuclear spins, and assume an external field B_0 much larger than the nuclear dipole fields B'. Then the spin of an implanted muon will rotate under the influence of B_0 and B' with

$$\omega(t) = \omega_0 + \omega'(t) \quad , \tag{3.14}$$

where ω' depends explicitly on time if the muon changes its environment due to diffusion. Different muons will experience different B' and the decay of the phase coherence is obtained from an ensemble average over all implanted muons

$$P(t) = \left\langle \exp\left[i \int_0^t dt' \omega'(t') \right] \right\rangle \quad . \tag{3.15}$$

For uncorrelated ω',P(t) is a Markov process. Usually (3.15) is evaluated under the additional assumption of a Gaussian process, where only the second-order cumulant of ω' is taken into account. Neglect of higher correlations leads to an overestimation of the motional narrowing, but proved itself as a good working hypothesis. For a Gaussian-Markovian process the second-order cumulant of ω' is given by:

$$\langle \omega'(t)\omega'(0) \rangle = 2\sigma^2 \exp(-t/\tau_c) \quad , \tag{3.16}$$

where τ_c is a field correlation time. This approach yields the well-known expression (3.9), which interpolates between a Gaussian line shape for a static muon and an exponential decay in the case of motional narrowing.

Here, we follow the approach by KEHR et al. /3.33/ and use a random walk description of the polarization decay, which is more adequate for jump diffusion and allows a straightforward extension toward diffusion with traps. We assume that the muon takes one frequency ω' for a mean time τ_c, then a second frequency ω" again for a mean time τ_c and so on. The frequencies ω', ω" ... are uncorrelated and distributed according to a Gauss law with a second moment $2\sigma^2$. This "strong collision" model is also a Markov process, but for finite τ_c the Gaussian assumption is no longer valid. The basic feature of the random-walk description is the separation of the polarization decay into contributions $R_\ell(t)$ of a fixed number ℓ of frequency changes

$$P(t) = \sum_{\ell=0}^{\infty} R_\ell(t) \quad . \tag{3.17}$$

The lowest-order terms have the following structure

$$R_0(t) = P_0(t)\exp(-t/\tau_0) \quad , \tag{3.18}$$

where $P_0(t) = \exp(-\sigma^2 t^2)$ is the static depolarization function and $\exp(-t/\tau_c)$ is the probability that the muon has performed no jump until t, and

$$R_1(t) = \int_0^t dt' \exp\left(- \frac{(t-t')}{\tau_c}\right) P_0(t-t') \frac{1}{\tau_c} \exp(-t'/\tau_c)\, P_0(t') \tag{3.19}$$

where $R_1(t)$ describes the polarization decay of muons, which have stayed until t' at one site. Then, at t' a transition to another site occurred [its probability is given by $(1/\tau_c)\exp(-t'/\tau_c)$]. Thereafter, they depolarize at the second site (probability: $\exp[-(t-t')/\tau_c]$). All possible contributions are obtained by integration over t'. The higher-order terms are similar convolutions including larger number of site changes. The whole series can be generated from an integral equation:

$$P(t) = R_0(t) + \frac{1}{\tau_c}\int_0^t dt'\, R_0(t-t')\, P(t') \quad . \tag{3.20}$$

The integral equation can be solved easily by Laplace transformation and P(t) is obtained by numerical inversion. A typical result is shown in Fig.3.7 and compared to the corresponding Gaussian-Markovian decay function which always lies below the random-walk result. This behavior is easily understood from the nature of the Gaussian assumption which neglects the correlations contained in the higher-order cumulants. Hence a more rapid decay results. The damping parameter Λ defined as

$$P(t = \Lambda^{-1}) = 1/e \tag{3.21}$$

is always smaller for the random-walk model compared to the Gaussian-Markovian model. The maximum difference is about 8% and the Λ of both theories become equal both for large and for small $\sigma\tau_c$. This is demonstrated in Fig.3.8 where the normalized damping parameters Λ/σ as a function of $(\sigma\tau_c)^{-1}$ are compared for both theories. Figure 3.8 also constitutes a typical temperature dependence expected for phonon-assisted diffusion. For low temperatures or long correlation times $(\sigma\tau_c)^{-1} < 0.1$ the muon is "frozen" in. At higher temperatures the muon diffusion accelerates and comes into the "time window" of a μSR experiment $0.1 < (\sigma\tau_c)^{-1} < 10$ where μ^+ diffusion can be observed. At even higher T the motional narrowing becomes complete and no further

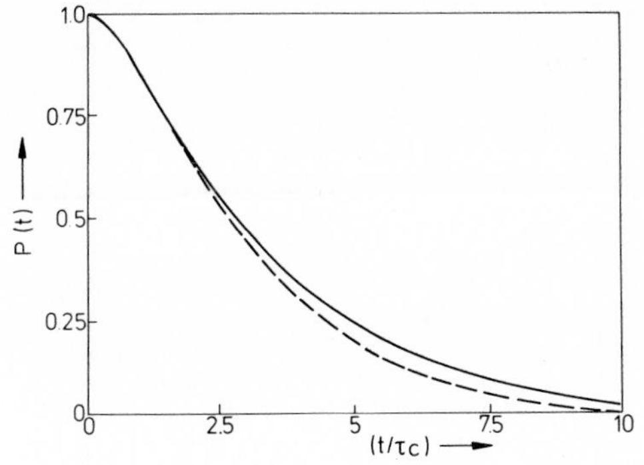

Fig.3.7. Polarization decay of the Markovian random-walk model (———) and of the Gaussian-Markovian theory (-----) for $(\sigma\tau_c)^2 = 0.2$

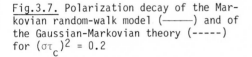

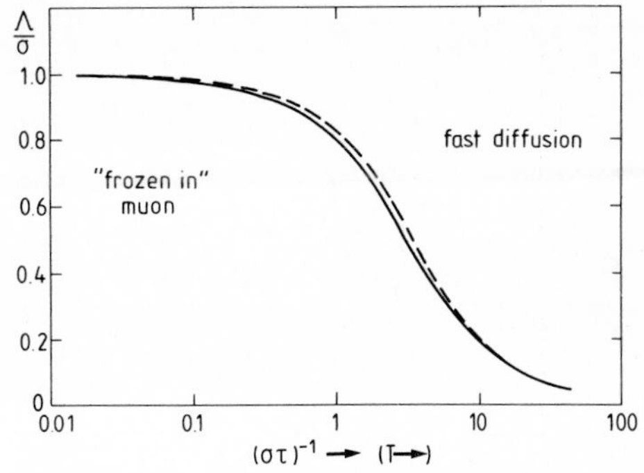

Fig.3.8. Comparison of the normalized damping parameter Λ/σ for the Markovian random-walk model (———) and the Gaussian-Markovian theory (-----) as a function of $(\sigma\tau_c)^{-1}$

information can be obtained. Other limitations are the lifetime of the muon – τ_c values longer than 10 τ_μ are hardly observable – and the time resolution of the electronics – Λ values larger than 10...50 μs^{-1} cannot be resolved.

The above treatment of line narrowing was restricted explicitly to the case of jump diffusion. Some considerations based on the work of McMULLEN and ZAREMBA /3.34/ which extend these ideas to coherent diffusion are presented in App.4.

3.5.2 Muon Diffusion – A Short History

The first muon diffusion experiments were undertaken about ten years ago by GUREVICH et al. /3.36/ on Cu. Since then a huge amount of data has been compiled and the course of our understanding of muon diffusion has taken some rather unexpected turns. In the first part of this section we present some of the "old" data. Thereafter in the second part we shall witness the path to the present knowledge on muon diffusion for the case of the bcc metals. In particular we shall concentrate on Nb.

Early results

Figure 3.9 presents the first μSR data on Cu which seemed to be the "text-book" case for a long time, since they nicely follow the theoretical expectation. From a low-temperature static linewidth the damping parameter starts to decrease around 80 K and falls off monotonically until it reaches complete motional narrowing around 300 K. The data have been evaluated using (3.9), and the resulting correlation times $1/\tau_c$ were presented in form of an Arrhenius plot with an activation energy of 47 meV. Later on GREBINNIK et al. /3.37/ complemented these data by an extension of the tem-

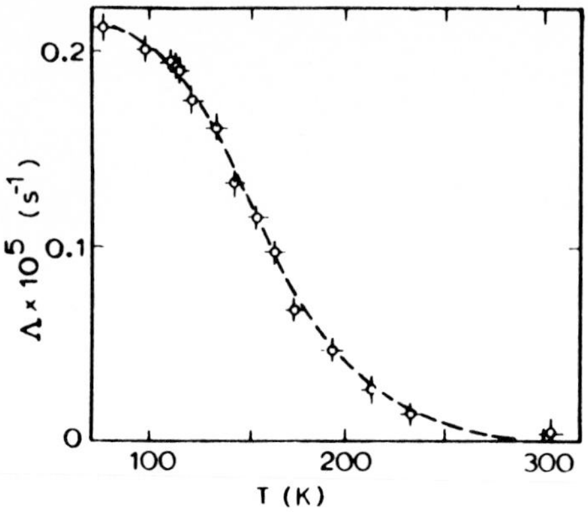

Fig.3.9. Muon depolarization data obtained from Cu as a function of temperature /3.36/

perature range and the inclusion of single crystal measurements. In terms of an Arrhenius relation the resulting correlation times obey the equation

$$1/\tau_c = 10^{7.61(4)} \cdot \exp(-48.4(1.5)\ [\text{meV}]/kT) \quad . \tag{3.22}$$

From a comparison with H diffusion data yielding a prefactor of $(1/\tau)_0 = 10^{+14}\ s^{-1}$ and an activation energy of 400 meV /3.38/, the authors concluded subbarrier motion for the muon which can account for both: the strongly reduced prefactor — no longer determined by an attempt frequency — and the drastically smaller activation energy — the muon does not climb the high barrier but tunnels through it. Finally, TEICHLER /2.16/ analyzed the data in terms of small-polaron theory using the Condon approximation and found E_a = 75 meV. The double force tensor $P = \psi a$ derived from E_a came out to P = 2.65 eV as already quoted. Applying (2.3) the self-trapping energy of the muon in Cu is about 150 meV. For the tunneling matrix element J, the rather small value of $J \cong 18$ μeV could be evaluated. Small deviations from the predictions of the small-polaron theory above 250 K were explained by contributions from above ground state transitions which were not included in the calculations. Summarizing, these Cu results are in very good agreement with the idea of small-polaron hopping. The qualitative difference between the behavior of the H and the muon in Cu is the difference between tunneling transitions from ground states in the case of the μ^+ and close to classical over-barrier jumping or lattice-activated tunneling (Sects. 2.3,4.1) in the case of the H.

New results on pure Cu in the temperature range 0.03 K $\leqq$ T $\leqq$ 5 K, however, show that μ^+ diffusion in Cu does not remain the textbook case it appeared to be for a long time /3.39/. Figure 3.10 displays the observed linewidth pattern which exhibits a low-T plateau at about 60% of the σ level at the higher temperature plateau. This behavior is not understood presently and several possible explanations have been put

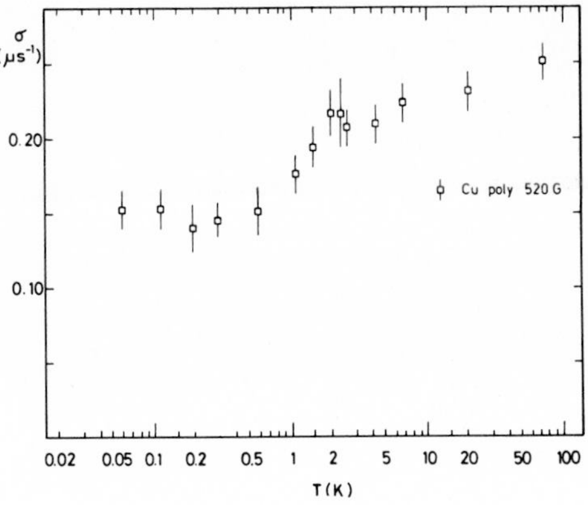

Fig.3.10. Depolarization parameter σ for polycrystalline Cu as a function of temperature /3.39/

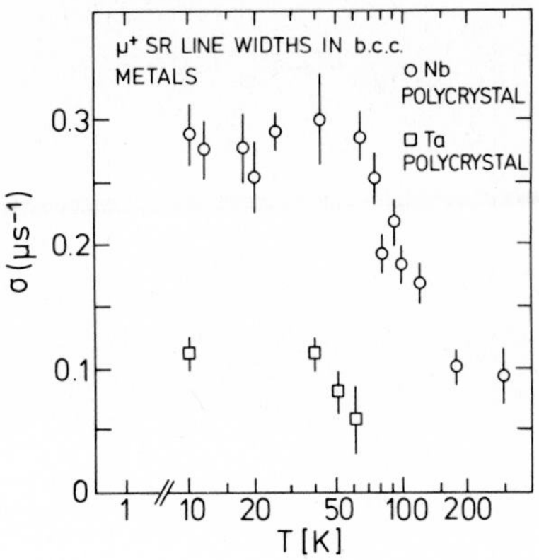

Fig.3.11. Early results on the temperature dependence of linewidths σ in Nb and Ta /3.30/

forward: (1) partial delocalization of the muon wave function below 2 K which would account for the smaller linewidth, (2) trapping over the entire temperature region, which implies that in sufficiently pure Cu, diffusion would be extremely fast, resulting in zero depolarization. Finally, also the possible influence of the random isotope mixture on the muon zero point levels has been discussed /3.40/ but conclusive experimental results are not available up to now.

Early results on the bcc metals Nb and Ta /3.30/ are shown in Fig.3.11. They look very similar to the Cu data showing a plateau region up to some 10 K followed by a more or less continuous decrease of σ with increasing temperature. A comparison of the muon jump rates with the corresponding values for H and D measured in the same temperature range revealed the puzzling result that muon diffusion comes out to be orders of magnitude slower than H diffusion in the same material. An exception from this more or less general behavior seemed to be the case of Al, where for all temperatures above 1 K complete motional narrowing had been observed /3.15/.

Muon diffusion in bcc metals

Further development of muon diffusion results, however, advises caution not to relate line narrowing data prematurely to intrinsic muon diffusion. In particular, the extreme sensitivity of muons toward impurities and defects gradually came into focus. Taking niobium as an example, this evolution will be followed up. Figure 3.12 presents results for the damping parameter Λ obtained from muons diffusing in Nb with different controlled amounts of interstitial impurities /3.41/ [Sample I has 3700 ppm N; Sample II, 60 ppm N and O; Sample III, 15 ppm N and O (in addition all sam-

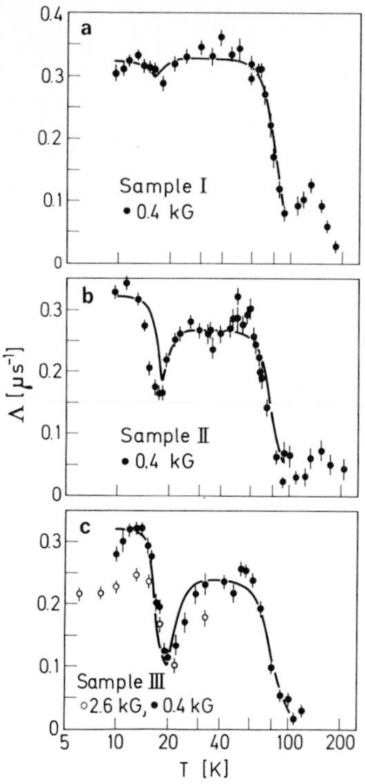

Fig.3.12a-c. The depolarization parameter Λ as a function of temperature for muons in Nb with different controlled amounts of impurities

ples contained about 100 ppm substitutional impurities, mainly Ta)]. While the data from the most impure sample (I) show almost the same temperature behavior as the Cu data, a characteristic dip appears in $\Lambda(T)$ at about 18 K for the more pure samples.

A similar dip phenomenon was first observed by GREBINNIK et al. /3.42/ for muon diffusion in Bi and was associated with a transition from incoherent diffusion at temperatures above the dip to quantum diffusion below the dip. The Nb data, however, show that the occurrence of the intermediate minimum in linewidth which deepens with increasing purity of the sample has to be related to the presence of impurities.

The data have been analyzed quantitatively with the two-state model of Sect.2.6, the consequences of which for muon depolarization will be outlined in the next paragraph. Here we give a qualitative description. Assuming that at low T the muons are localized at random in fixed lattice positions, the first drop in linewidth can be ascribed to the onset of free diffusion not influenced by traps. With increasing temperature the μ^+ mobility increases and the muons are able to reach the traps caused by the interstitial N and O atoms. An increase of the damping above 20 K results. The depth of the dip phenomenon is naturally related to the interstitial impurity concentration, since the trapping probability decreases with trap concentration. In the broad plateau region between 30 K and 60 K, the muons experience the dipolar fields in the traps for most of their lifetime. The sharp drop of damping above 60 K indicates the beginning of escape processes from the traps. The motional narrowing is then caused by repeated capture and release processes. The dissociation energy from a N trap has been evaluated as $E_a^{\mu} = 48$ meV and should be compared with the corresponding value for H, $E_a^H = 167$ meV obtained from a quasielastic neutron scattering experiment /2.39/.

Results comparable to those for Sample III have also been observed by BIRNBAUM et al. /3.31/, who studied a purified Nb single crystal and gave a similar qualitative description without attempting a quantitative analysis.

We note that the observed dip structure is a consequence of the nonequilibrium initial conditions for the muon: the muons are thermalized at random position in the lattice not obeying the Boltzmann statistics. A similar experiment on H in Nb would not reveal any intermediate minimum in linewidth, since at low T all protons would sit in the deepest traps and not be released until thermal activation out of the trap is possible.

Already the quantitative analysis of the above muon data in Nb had difficulties to explain the relation between the capture rate at the impurities $1/\tau_1$ and the jump rate for free diffusion $1/\tau$. The ratio of both came out as too small to be explained by diffusion limited trapping and fostered speculations on drift motion toward the traps. Measurements on the impact of substitutional impurities like V on the muon diffusion in Nb /3.43/ soon raised additional doubts whether the assumption of randomly localized muons in Nb below 15 K is valid. As Fig.3.13 shows, V was found to shift the "onset of diffusion" toward higher temperatures in contradiction to the idea of "free diffusion" which is not affected by impurities in this region.

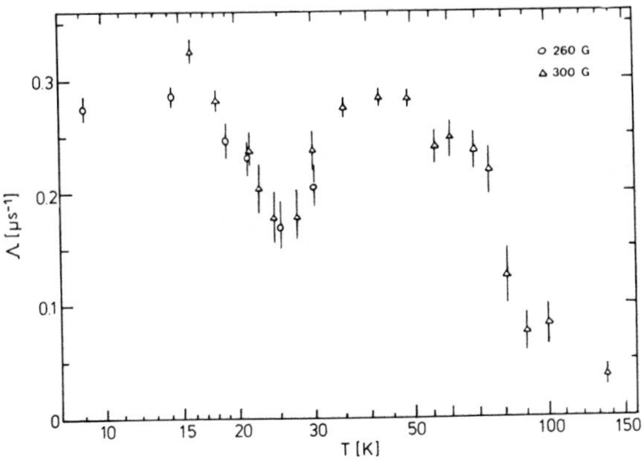

Fig.3.13. Muon depolarization data on a $NbV_{500\ ppm}$ alloy indicate a shift of the "onset of diffusion" toward higher T

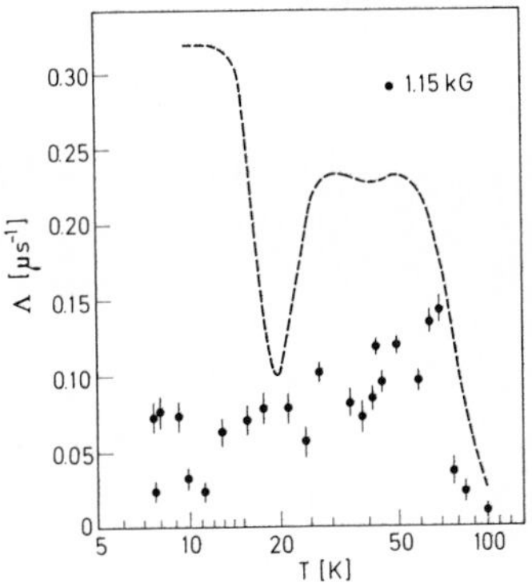

Fig.3.14. Λ(T) for ultra pure Nb. The dashed line shows the results on the purest sample, III, from Fig.3.12

Later experiments on ultra pure Nb, where most of the substitutional impurities were removed by fused salt electrolysis (remaining substitutional impurity concentration 2 - 3 ppm, mainly Ta) and where the interstitial impurities were reduced to 1 - 2 ppm by ultra high vacuum zone melting, revealed the astonishing result of high muon mobility in the overall temperature region /3.43/. Figure 3.14 presents the results obtained from the ultra pure material and compares them with the outcome of the foremost purest sample, III. Nearly all structures including the low-temperature plateau observed in Samples I to III vanish and only a small reminder on the N trapping peak remains. Indications of such behavior have also been reported by BROWNE et al. /3.44/ and by METZ et al. /3.45/. Both groups of authors found a significant drop of the low- and high-T plateaus in going from pure Nb to ultra pure Nb, where the Ta impurities are removed. However, presumably due to the higher interstitial impurity content in their samples, the effect was not as dramatic as shown in Fig.3.14.

Very recently these studies have been extended to a larger temperature range, even more pure samples and pure samples doped with controlled amounts of Ta (50 ppm) or N (15 ppm) /3.46/. The results are summarized in Fig.3.15. Figure 3.15a presents the μ^+ depolarization in pure Nb over three orders of magnitude in temperature commencing at 0.1 K. Three features are noteworthy: (i) between 5 and 50 K the damping σ stays on a constant low level $\sigma \cong 0.1\ \mu s^{-1}$; (ii) toward lower T the damping increases and reaches a second plateau which extends to the lowest temperatures. The plateau value of σ is not related to the static muon linewidth in Nb;

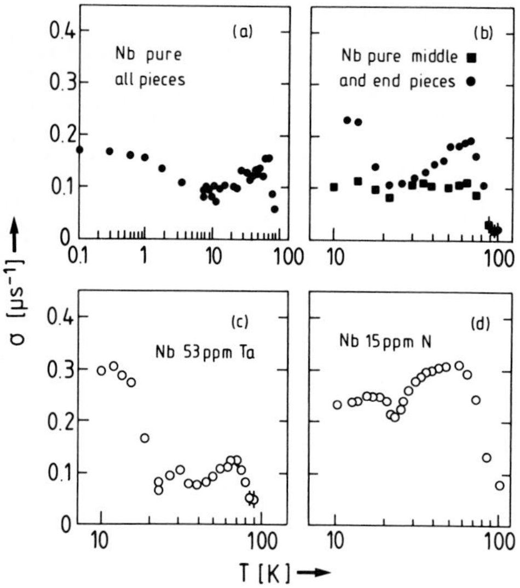

Fig.3.15a-d. Muon depolarization in ultra pure Nb (a,b) and in ultra pure Nb doped with Ta (c) and N (d)

(iii) between 50 and 100 K the typical N or O trapping peak arises. Its height is of similar magnitude to the level of the low-T plateau.

The ultra pure Nb sample used in the experiments of Figs.3.14 and 3.15a consisted of 19 rods, originating from 5 different single crystals. Figure 3.15b displays measurements on two different selections of pieces from the purest single crystal (resistivity ratio 11600). The sample called "end pieces" was taken from the outer part, the one called "middle pieces" from the center of the single crystalline rod. The contents of impurities are naturally slightly higher in the outer parts due to the zone-refining process, but the influence on the muon depolarization is striking.

Figures 3.15c and d show the effect of doping the ultra pure Nb with selected impurities. Figure 3.15c presents the effect of 53 ppm Ta, which yield the same low-T plateau as observed before on Samples II and III (Fig.3.12), while only small indications of the N(O) trapping peak at higher temperature remain. Finally, Fig. 3.15d presents the effect of 15 ppm N which causes a plateau in linewidth below 30 K, slightly lower than the σ observed in the case of Ta doping, and a peak between 50 and 100 K.

The authors point out that the results in Fig.3.15 together with those of references /3.44,45/ cannot be explained by muon diffusion and trapping only, and suggest a consideration of delays to self-trapping (Sect.2.7). A precondition for significant delays is a sufficiently large ratio A of free bandwidth zJ_f and lattice frequencies (2.34). For Nb, A was estimated to about 1.8 ... 2.5 significantly larger than the estimate for Al: $A_{Al} \cong 1$. Thus, only a fraction α of the implanted muons can be anticipated to self-trap by thermal fluctuations. According to (2.36) this fraction would be temperature independent for $T < \Theta_D/2$. Another fraction β of the muons is expected to localize by static strains near impurities. For muons propagating in a metastable state with large free path ℓ, this process is capture controlled and can be described by (2.32). Consequently the corresponding rate is proportional to the impurity concentration and *independent* of temperature. In contrast to random localization by thermal strains, strain localization by impurities always implies subsequent trapping of the small polaron thus created, which can move only if the temperature is high enough to dissociate it from the trap. A third fraction γ might never get self-trapped and diffuses always in a metastable state.

These 3 different muon fractions are expected to exhibit very different depolarization behaviors as a function of temperature and their contribution to the observed linewidth is displayed schematically in Fig.3.16.

α: These muons form small-polaron states at random positions in the lattice and perform small-polaron diffusion. At low T coherent diffusion increasing with decreasing T is expected, while phonon-assisted processes should dominate at

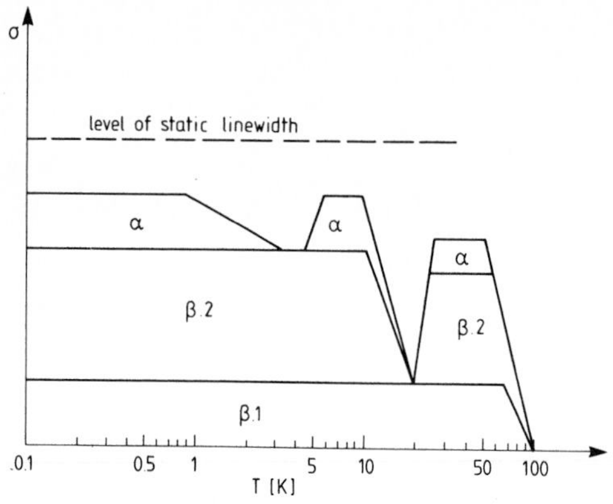

Fig.3.16. Schematic sketch of the linewidth contributions from the different muon fractions [(α) thermally self-trapped, (β.1) self-trapped by interstitial impurities, (β.2) self-trapped by substitutional impurities]

higher T. Thus, both at low as well as at high temperatures diffusion is fast enough to reach trapping centers, where the muons are depolarized.

β.1: If the traps which induce self-trapping are *deep* (as expected for interstitial impurities) this fraction of muons is released only at high T and gives rise to temperature-independent depolarization below the release temperature, which is about 80 K for N and O impurities.

β.2: For more shallow traps (like those provided by substitutional impurities) the muons are released to lower T and subsequently diffuse as small polarons. At higher T these muons, initially self-trapped by static strains, behave like fraction α and may be caught by deeper traps in a diffusion-controlled way.

γ: The muons in the metastable propagating state are expected to give no contribution to σ at any temperature, since the motional narrowing should be complete.

The predicted characteristic temperature dependencies can now be compared with the experimental results of Fig.3.15. The data in Fig.3.15a correspond to the case α + β.1. The temperature-independent plateau stems from muons localized at the N or O impurities, while the increase in depolarization below 1 K and above 50 K originates from the small-polaron fraction α. The rest of the muons stay in the metastable state and create no depolarization at any temperature. In Fig.3.15b the data from the end pieces reflect the situation α + β.2 (main contribution) + β.1. The nitrogen concentration is small such that only a fraction of β.2 reaches the deep traps around 70 K. The σ values from the middle pieces on the other hand indicate a very small amount of shallow traps, the plateau stems from a fraction β.1. In this connection also the data from references /3.44,45/ can be understood as resulting mainly from strain localization in deep traps (β.1). Figure 3.15c corresponds to α + β.2 and a small admixture of β.1. Figure 3.15c demonstrates that at T < 10 K a plateau in σ is observed which corresponds to capture-controlled localization at the Ta;

the small increase around 12 K stems from the fraction α reaching the Ta trap (Fig. 3.12c). Finally Fig.3.15d corresponds to $\alpha + \beta.2$: most of the muons are self-trapped at the N atoms up to 80 K.

Summarizing, the present results on Nb show muon behavior characterized by the following features: (i) muon diffusion in pure Nb is extremely fast and can be observed via impurity trapping only; (ii) small amounts of impurities have a crucial impact on the observed depolarization parameter σ. While substitutional impurities like Ta or V release the muon around 20 K, interstitial impurities like O or N keep the muon trapped up to about 80 K; (iii) small-polaron formation does not occur instantaneously but may be induced either by thermal fluctuations or by lattice strains due to impurities; (iv) there is experimental evidence for muons which do not form a small-polaron state during their nuclear lifetime, but which always stay in a metastable propagating state.

The development of our knowledge on muon diffusion in the other two commonly investigated nonmagnetic bcc metals Ta and V took a similar course as in Nb. After the early Cu-like results on Ta, METZ et al. /3.45/ observed a dip phenomenon also on purified Ta which they interpreted in analogy to the Russian interpretation /3.42/ of the Bi results by an onset of quantum diffusion. In view of the more advanced Nb results, however, it is almost certain that the observed local minimum in linewidth is again an impurity effect.

Concerning V, the starting point again was a Cu-like temperature dependence of the damping parameter. Further progress, however, showed some qualitative differences to the other two Vb metals and demonstrates again how impurity effects can lead to severe misinterpretations of diffusion data. FIORY et al. /3.26/ reported a linear temperature dependence of the muom jump rate $1/\tau_c$ obtained from a V sample with a nominal purity of 99.2%. It appeared obvious to relate such a linear T dependence to a one-phonon mechanism. On the basis of Sussmann's theory which disregards small-polaron effects and does not correctly predict the explicit dependence of the jump rate on the strain splitting ΔE (2.15,16), the authors concluded that one-phonon process is prominent in V up to 100 K.

HEFFNER et al. /3.27/ investigated V specimens with different degrees of purity. Their results for the observed damping λ taken from an exponential fit $[P(t) \sim \exp(-\lambda t)]$ are displayed in Fig.3.17 together with the data of FIORY et al. The purest V which still contained about 50 ppm interstitial and 50 ppm substitutional impurities reveals a sharp peak around 100 K and a slowly increasing depolarization rate below 30 K. Doping with 500 ppm O changes the linewidth profile completely. Astonishingly, O doping does not result in broadening and increasing the 100 K peak, but enlarges the damping towards lower T with the outcome of a Cu-like T behavior. Thus, the 100 K peak does not seem to be associated with the presence of O. A fur-

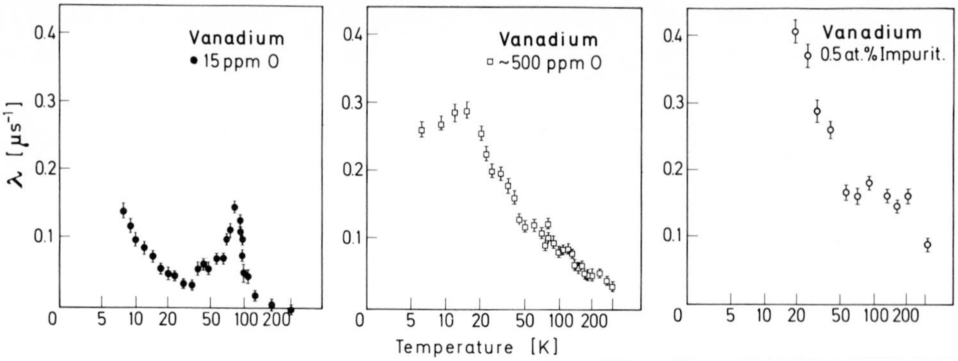

Fig.3.17. Exponential μ^+ depolarization rate $[P(t) = \exp(-\lambda t)]$ for three vanadium samples of different impurity contents, as a function of temperature, right from /3.26/, others from /3.27/

ther increase of the impurity concentration as present in FIORY's sample (Fig.3.17c) causes an even steeper growth of depolarization below 30 K. An explanation of the systematic increase of σ or decrease of muon mobility with growing impurity content in terms of one-phonon processes fails. According to (2.21) the jump rate is proportional to $(\Delta E)^2$, and an increase of mobility with enhanced impurity content would be expected.

Again, the observed depolarization data appear to be extremely sensitive to trapping at impurities, and as in the case of Nb it is very likely that all structure in linewidth found so far is not related to intrinsic muon diffusion. To what extent localization processes may play a role in V is not clear. Compared to Nb the even larger value of $A = zJ_f/\hbar\omega_D$ (3.34) for muons in V, however, indicates that delay to self-trapping may also be important for V.

3.6 Muon Diffusion and Trapping – The Case of Al

Compared to the development of experimental evidence on all other metals, the outcome of muon diffusion experiments on Al, which is easily purified to a 1 ppm level, took an inverse course. At the beginning stood the exceptional result of high muon mobility down to the lowest accessible temperatures /3.15 (1 K)/ (presently 25 mK /3.39/). Soon it was noticed that small quantities of impurities like Cu /3.47,48/ or Mn /3.19/ or intrinsic defects from deformation /3.48/ or irradiation /3.50/ induce muon localization at low temperatures. While GREBINNIK et al. /3.48/ as well as HARTMANN et al. /3.19/ stressed the influence of random strainfields created by the impurities on the persistence of coherent diffusion, KOSSLER et al. /3.47/ and

DORENBURG et al. /3.50/ emphasized the aspect of impurity trapping. In particular in the case of Cu impurities in Al, KOSSLER et al. /3.47/ interpreted their data as resulting from trapping on up to five (!) different types of Cu microclusters. Also the influence of impurities like Ag, Mg, Si, Zn, Li was reported /3.49,51,52/. Very recently KEHR et al. /3.52,2.27/ investigated systematically the muon localization-delocalization phenomena taking place in Al doped with substitutional impurities at concentrations of 100 ppm and below. The emphasis of these series of experiments was placed on the underlying muon diffusion mechanisms which were studied indirectly through the localization and trapping mechanisms. In this section we shall mainly concentrate on these results which up to now yielded the deepest insight into muon diffusion properties in metals. Since all information is obtained via the influence of impurities we shall commence with an outline of the theory of muon depolarization for muon diffusion in the presence of traps as given by KEHR et al. /3.33,53/.

3.6.1 Muon Depolarization for Diffusion in the Presence of Traps

The calculation of muon depolarization for diffusion in the presence of traps is a straightforward extension of the random-walk description of motional narrowing through jump diffusion (Sect.3.5.1). We employ the two-state model introduced in Sect.2.6, where the muon diffuses either in a free state with a mean time of stay τ_1 or is caught by a trap for an average time τ_0. Such a two-state model is an average representation of the actual capture and release processes and is of similar structure to the elementary processes in the strong-collision model. Let $P_f(t)$ and $P_t(t)$ be the polarization decays in the free or the trapped state, respectively. For P_f we take the polarization decay in the undisturbed medium as defined by (3.26). In the simplest approximation $P_t(t)$ is given by a Gaussian decay $P_t(t) = \exp(-\sigma_t^2 t^2)$ which assumes an immobile muon in the trapped state and σ_t^2 is the second moment of the frequency distribution in the trap. Finally, $Q_f(t)$ and $Q_t(t)$ are the polarization decays for muons disintegrating in the free or trapped state, irrespective of their previous history. For the sake of simplicity we assume $Q_f(0) = 1$ and $Q_t(0) = 0$ corresponding to muons always starting in the free state. This is a good approximation for random implantation and low impurity concentration. Then, in analogy to (3.20) Q_f and Q_t fulfill the integral equations

$$Q_f(t) = P_f(t)\exp(-t/\tau_1) + \frac{1}{\tau_0}\int_0^t dt'\ P_f(t-t')\exp[-(t-t')/\tau_0]\ Q_t(t')$$

$$Q_t(t) = \frac{1}{\tau_1}\int_0^t dt'\ P_t(t-t')\exp[-(t-t')/\tau_0]\ Q_f(t') \quad . \qquad (3.23)$$

Its iteration generates series which now represent the different possibilities for depolarization in terms of contributions from fixed numbers of state changes (no trapping until t, or one trapping process, or two, etc.). The integral equations can be solved analytically by Laplace transformation. Since the experiment cannot distinguish between positrons originating from muons decaying in the trap or in the free state, the muon depolarization Q(t) is finally given by the sum of $Q_f(t)$ and $Q_t(t)$ which can be obtained by numerical inversion of the Laplace transform.

As has been shown in /3.33/, the integral equations can be approximated by a system of differential equations, which include changes between free and trapped states and reduce to Abragam type expressions (3.9) in the limit $\tau_1, \tau_0 \to \infty$:

$$\frac{dQ_f}{dt} = 2\sigma^2\tau_c\left[\exp(-t/\tau_c)-1\right] Q_f - \frac{1}{\tau_1} Q_f + \frac{1}{\tau_0} Q_t$$

$$\frac{dQ_t}{dt} = 2\sigma_t^2\tau_0\left[\exp(-t/\tau_0)-1\right] Q_t - \frac{1}{\tau_0} Q_t + \frac{1}{\tau_1} Q_f \quad . \tag{3.24}$$

The resulting depolarization function is a good approximation for linewidth as well as line shape to the solution of the integral equation and can be utilized much easier for the purpose of fitting experimental data. It has been applied, for example, to analyze the muon data obtained from Nb with different amounts of impurities, shown in Sect.3.5.2. The solid lines in Fig.3.12 represent the result.

Other trapping models which start from the muon correlation function in a Gaussian-Markovian approach or which do not allow for escape processes are also in the literature but will not be discussed here /3.54,55/.

3.6.2 Muon Diffusion in Aluminium

As mentioned already, a strong motional narrowing of the μSR signal occurs in the purest Al samples even at temperatures as low as 30 mK. Thus, diffusion can be studied only indirectly via trapping mechanisms which at least for diffusion-limited trapping yield direct information on the diffusion properties. Extremely well characterized samples are a very important precondition of such experiments, since, as we have noticed already for the case of Nb, impurities on a ppm level may affect the muon strongly. Here we are concerned with recent measurements by KEHR et al. /2.27/ on pure Al doped with controlled amounts of Mn, Mg, Ag and Li impurities, which provide widely different long-range effects on the host lattice: Mn and Mg impurities produce large volume changes per atom and according to (2.20) create long-range energy fluctuation in the host; Ag and Li on the other hand fit very well into the

Al lattice and besides some short-range disturbances leave the host lattice intact. The experiments on impurity trapping revealed detailed information on muon diffusion behavior between 30 mK and 50 K, and were complemented by studies on vacancy trapping by HERLACH /3.56/ which extend the temperature range to higher T. Both sets of data provide a firm basis for the discussion of the different diffusion mechanisms outlined in Sect.2.

Figure 3.18 presents an overview on the results obtained from Al doped with different amounts of Mn impurities also including data from pure Al. The temperature dependence of σ suggests a division of the observed data into two different regions: (i) a low-T regime ($T < 3$ K), where σ is roughly proportional to $\ell n(1/T)$ and (ii) a high-T region which is characterized by peaks in the depolarization rate.

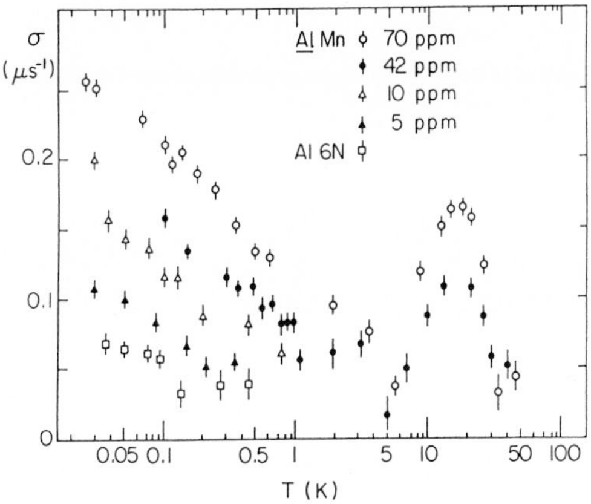

Fig.3.18. Gaussian depolarization parameter σ for Al and AlMn polycrystalline samples. The external field is B_0 = 520 G for $AlMn_{42\ ppm}$ and 150 G for the others

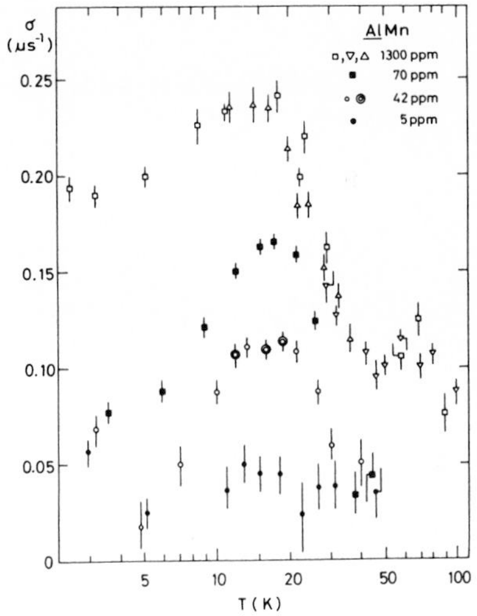

Fig.3.19. The trapping peak region around 15 K for AlMn samples of various concentrations. Some of the samples have been measured at more than one magnetic field, which is indicated by different symbols. The 42 ppm points are at higher field than the others except for the three points marked with a double ring. These were measured at 130 G and scaled down 15% in order to accommodate the field dependence of linewidth

Muon diffusion above 3 K — phonon-assisted motion

The temperature and concentration dependence of σ above 3 K exhibits the typical features of capture and escape processes from traps. Figure 3.19 displays this region in more detail for AlMn with various Mn concentrations. The linewidth in all $AlMn_x$ samples exhibit a pronounced maximum around 17 K. The height of the peaks depends on the impurity concentration n_{Mn}, while their position is not affected by it. For the highest concentration n_{Mn} = 1300 ppm clear indications of saturation are seen, i.e., around 17 K practically all the muons stay in traps during the time of observation. In addition a secondary peak evolves around 65 K. Such extra peaks seem to be characteristic for strongly doped materials and have also been found, e.g., in Nb doped with 3700 ppm N (Fig.3.12) or in other heavily doped Al compounds /3.49/ and presumably originate from impurity clusters /3.47/. Measurements on Al samples doped with other kinds of impurities are shown in Fig.3.20. The maxima of linewidth occur at different temperatures for each impurity. While for <u>Al</u>Mn and <u>Al</u>Ag the muon can escape from the trap around 20 K, this temperature is shifted to around 50 K for <u>Al</u>Mn and <u>Al</u>Li.

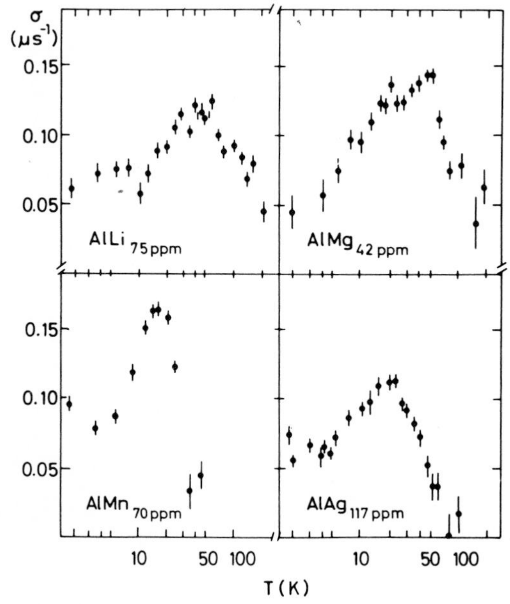

Fig.3.20. Gaussian depolarization parameter σ for Al with various doping elements for T = 2 - 100 K

Quantitatively the data have been analyzed with the two-state model described in Sects.2.6 and 3.6.1. Thereby, considering the complete motional narrowing in pure Al, fast diffusion inbetween the traps was supposed and depolarization in the 'free state' was neglected. This corresponds to $\tau_c \to 0$ and simplifies the differential equations (3.24) considerably. For the temperature dependence of the escape rate an Arrhenius-type behavior $1/\tau_0 = \Gamma_0 \exp(-E_0/kT)$ was assumed, while the trapping rate was set proportional to a power of T $1/\tau_1 = \Gamma_1 T^{\alpha}$ reflecting the theoretically predicted behavior.

For $\underline{\text{Al}}\text{Mn}_x$ the following results were obtained. (i) The trapping rate $1/\tau_1$ was found to be proportional to the Mn concentration. Thus trapping occurs on single impurities (exception $\underline{\text{Al}}\text{Mn}_{1300}$). (ii) The trapping rate is only weakly temperature dependent, i.e., $1/\tau_1 \sim T^\beta$ ($\beta < 1$). (iii) The dissociation energy from a Mn trap was found to be 10 meV compared to 48 meV for a N trap in Nb.

In order to illustrate the results, we mention as an example the values of τ_1 and τ_0 at the peak temperature (17 K) in $\underline{\text{Al}}\text{Mn}_{70\ \text{ppm}}$: τ_1 = 3.3 μs and τ_0 = 40 μs. Thus, not every muon is caught by a trap during its lifetime τ_μ = 2.2 μs in this sample at 17 K; if however, trapping occurs, the muon can hardly escape. Consequently, the peak height is directly related to the fraction of muons trapped.

It is interesting to note that the correlation between the trapping capability of a particular impurity and the lattice expansion it creates (Table 3.3) is only weak. The escape rates do not seem to be correlated at all. Or, in other words, the local distortion around the impurity which determines the trapping behavior cannot be estimated from the long-range strain fields. This is in agreement with recent diffuse neutron scattering results on $\underline{\text{Al}}\text{Li}$ which show short-range atomic displacements even with a net volume dilation close to zero /3.57/.

Table 3.3. Lattice expansion of Al due to substitutional impurities deduced from /2.32/

Impurity	Mn	Mg	Li	Ag
$\Delta a/a$ for 1 at.%	$-1.5\cdot10^{-3}$	$0.99\cdot10^{-3}$	$-1.1\cdot10^{-4}$	10^{-5}
ΔV Å^3	-7.42	4.9	-0.54	0.05

Concerning the underlying diffusion mechanism, the magnitude and temperature dependence of the trapping rate $1/\tau_1$ is of prime interest. According to (2.31) the diffusion coefficient is directly proportional to $1/\tau_1$. For the sake of simplicity we assume a trapping radius r_t = a. Applying (2.31), the diffusion coefficient ($3 < T < 20$ K)

$$D \cong 1.3\cdot10^{-8}\ T^{0.9}\,[\text{cm}^2/\text{s}] \qquad (3.25)$$

is obtained from the $\underline{\text{Al}}\text{Mn}$ data. The nearly linear temperature dependence of D is in obvious disagreement with the prediction of a T^7 power law due to two-phonon processes (2.9). On the other hand, it is in qualitative accordance with the rate obtained for one-phonon processes (2.16).

One-phonon processes are possible either if strain fields disturb the translational symmetry — the jump rate then is proportional to the mean square disturbance $<\Delta E^2>$ (2.16) — or if jumps occur between inequivalent sites [e.g., O-T-O jump sequences (2.17)]. For the strain-induced part, an increase of the jump rate with growing impurity concentration is expected — in Sect.2.4 a $n_t^{8/3}$ dependence was estimated. However, the experimental trapping rate showed only the linear concentration dependence related to the trapping kinetics; the resulting diffusion coefficient does not depend on n_t. In addition, impurities with strong and weak long-range distortion fields should be accompanied by very different diffusion coefficients. In contrast to these predictions muons in AlAg and AlMn which are subject to very different long-range strain disturbances exhibit comparable trapping peaks at comparable impurity concentration.

There are two possibilities for jumps between inequivalent sites: (i) reorientational changes of the double force tensor during a jump [(2.17), $P_a \neq 0$] and (ii) jumps between crystallographically different sites. The cubic symmetry of both T and O sites in fcc lattices should always be accompanied by an isotropic double force tensor, and reorientation cannot play a role. Also processes beyond the Condon approximation, where one-phonon processes influence J directly, are not very likely for jumps between equivalent sites, since in fcc lattices the symmetry requirements are not fulfilled (App.2). The latter case applies, e.g., for jumps between O and T sites in a fcc structure. Such a process is independent of the long-range energy fluctuations ΔE, as long as the difference $\Delta\varepsilon$ between the O and T site is much larger than ΔE. It is linear in T for $kT >> \Delta\varepsilon$, which in our case requires values of $\Delta\varepsilon$ well below 1 meV. The observed shift of occupation from a T site around 15 K to an O site at lowest temperatures indicates a delicate balance between O and T sites in Al and thus, a small $\Delta\varepsilon$. However, it cannot be estimated presently how much the tetrahedral site occupation found at the Mn peak position around 15 K is influenced by the presence of the Mn impurity. One-phonon processes beyond the Condon approximation could also contribute to the jump rates between unlike sites. However, no estimate is available so far.

HERLACH /3.56/ reported muon diffusion results from vacancy trapping in electron-irradiated Al. Figure 3.21 presents results for the 1/e decay Λ observed for different vacancy concentrations. Here Λ increases with temperature toward a saturation value Γ_0^{1V} representing the static muon linewidth in a monovacancy. The starting concentration was $n_v = 4 \cdot 10^{-6}$, which was reduced in different steps by subsequent annealing. The observed increase of damping with temperature is due to diffusion-controlled trapping, where the muon mobility increases with temperature and the muons cannot escape the trap. The shift of the damping increase observed after annealing reflects the recovery of radiation damage as a function of the annealing

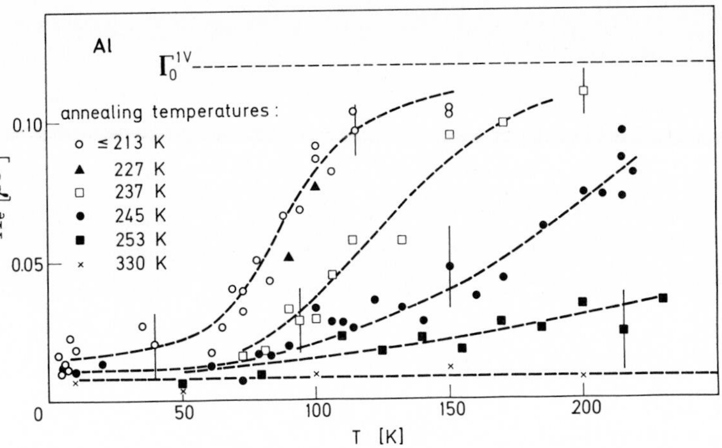

Fig.3.21. Muon depolarization Λ_e in electron-irradiated Al (vacancy concentration before first annealing $n_V = 4\cdot10^{-6}$)

temperature. The data are a good example for the extreme sensitivity of the muon toward lattice defects. Even vacancy concentrations below the 1 ppm level are easily detected with muons. The authors estimate a vacancy concentration of $7\cdot10^{-7}$ after annealing at 245 K, which still increases the muon depolarization strongly.

Again assuming diffusion-controlled trapping, METZ /3.58/ has analyzed the vacancy trapping data in order to extract the diffusion coefficient. The result is shown in Fig.3.22, where the obtained diffusion coefficients are plotted in an Arrhenius representation. The dashed line indicates a temperature dependence as pre-

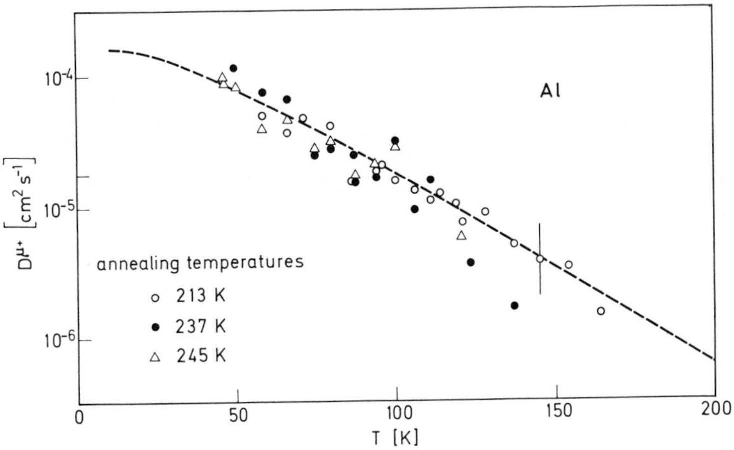

Fig.3.22. Temperature dependence of the muon diffusion coefficient derived from vacancy trapping data assuming diffusion-controlled trapping. (-----) prediction of (2.6)

dicted by small-polaron theory in the high-T limit (2.6). Above 60 K the diffusion coefficient follows the equation

$$D = 7\cdot10^{-3}\ T^{-1/2}\exp(-32\,[\mathrm{meV}]/kT)\ [\mathrm{cm}^2/\mathrm{s}] \qquad (3.26)$$

in good agreement with small-polaron theory. Using (2.6), the tunneling matrix element can be estimated to J = 2.3 meV. Considering the uncertainties of its evaluation, it should be taken as an order of magnitude estimate. But nevertheless it is two orders of magnitude larger than the value of 18 μeV found in Cu. The activation energy of 32 meV on the other hand is considerably lower than the 48 meV taken from an Arrhenius representation of the Cu results (3.22).

Utilizing the evaluated parameters J and E_a, we estimate whether (2.17) which describes the one-phonon process for jumps between inequivalent sites is able to reproduce the magnitude of the diffusion coefficient in the region $2 \leqq T \leqq 20$ K as given by (3.25). Inserting all elastic data and P = 1.3 eV from Sect.3.4, we obtain a diffusion coefficient

$$D \cong 3\cdot10^{-5}\ (\Delta\varepsilon\,[\mathrm{eV}])^2\ T\,[\mathrm{K}]\ [\mathrm{cm}^2/\mathrm{s}]$$

Allowing for energy differences of the order of $\Delta\varepsilon \leqq 10^{-3}$ eV, to get a linear T dependence, the resulting D is about 3 orders of magnitude smaller than the experimental result. Even in view of all the uncertainties involved, this discrepancy seems to be serious and may indicate the presence of other than one-phonon-transfer-processes which are not taken properly into account so far.

The results below 3 K – quantum diffusion

The characteristic feature of the depolarization data in Al doped with substitutional impurities below 3 K is a monotonic increase of the depolarization rate with decreasing temperature. Thereby at a given temperature σ increases with growing impurity concentration (Fig.3.18). While above 3 K different impurities cause trapping peaks characteristic for each kind of impurity, the low-T results for <u>Al</u>Mn, <u>Al</u>Li, <u>Al</u>Ag are remarkably similar. Figure 3.23 shows low-T data taken from Al doped with Ag, Li and Mn impurities and it is obvious that the functional form of the temperature dependence is identical for all investigated impurities. While at a given concentration Mn seems to have the largest impact, Ag appears to cause the smallest increase in depolarization. This may be related to the local lattice distortion around the impurity which is largest for manganese and smallest for silver.

The results of the data evaluation in terms of correlation times τ_c (3.9) are presented in Fig.3.24. Three features are noteworthy: (i) τ_c^{-1} follows simple power

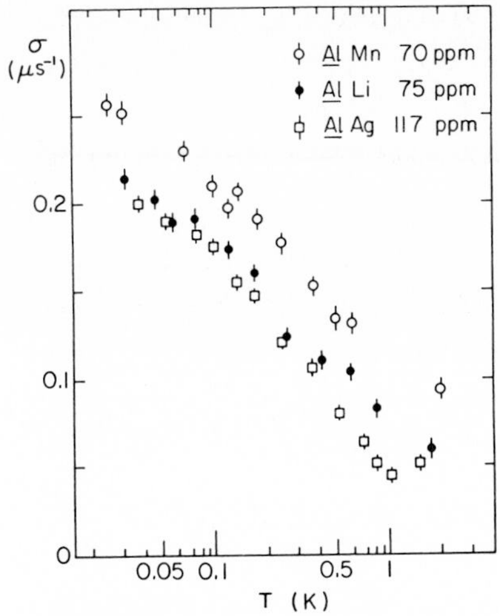

Fig.3.23

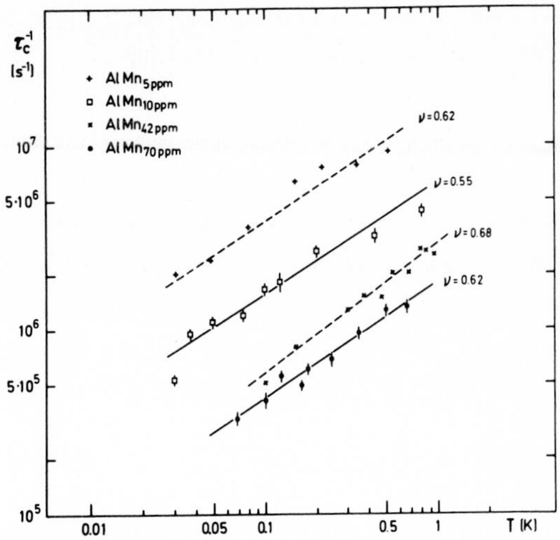

Fig.3.24

Fig.3.23. Depolarization parameters σ for <u>Al</u>Mn, <u>Al</u>Li and <u>Al</u>Ag samples below 2 K

Fig.3.24. Inverse correlation times τ_c^{-1} for various <u>Al</u>Mn samples as a function of T. The lines are individual fits to a T^ν dependence with the parameter ν given here. A global fit of the data gives $\nu = 0.60(4)$

laws $\tau_c^{-1} \sim T^\nu$ over more than one order of magnitude in temperature; (ii) τ_c^{-1} increases with increasing manganese concentration, the concentration and temperature dependencies can be expressed with good accuracy in a closed form

$$\tau_c^{-1} = 6.9(1.2)\cdot10^5 \ T[mK]^{0.60(4)} n_{Mn} \ [ppm]^{-0.76(4)} \ [s^{-1}] \quad ; \tag{3.27}$$

(iii) though causing very different long-range disturbances, Mn, Li and Ag have essentially the same effect on muon behavior. Firstly we consider whether these results can be understood by uniform muon diffusion disregarding trapping which occurs above 3 K:

(i) From the inverse concentration dependence of $\tau_c^{-1} \sim n_{Mn}^{0.76}$ it is evident immediately that the underlying motional process cannot be an incoherent one-phonon jump mechanism as described by (2.15-17). In the case of diffusion brought about by the long-range disturbances ΔE, a proportionality to $(\Delta E)^2$ should hold and τ_c^{-1} should *increase* with increasing impurity concentration and behave qualitatively different for impurities with long- and short-range distortion fields in contrast to the observations. For -O-T-O-T- jumps the rate should not depend on n_t.

(ii) Uniform coherent diffusion could account for the observed concentration dependence of the depolarization if impurity scattering is invoked as the dominant transport scattering mechanism. Applying (2.27,25) for $kT << zJ_{eff}$, the coherent diffusion coefficient should behave as $D_c \sim T^{1/2}\ c^{-1}$ very close to the experimental findings. However, considering the magnitudes of $1/\tau_d$ (2.28) and of the thermal velocity, D comes out many orders of magnitude too large and would cause complete motional narrowing over the whole T range. In addition, coherent diffusion limited by impurity scattering would require an abrupt change from incoherent hopping between localized states above 3 K to freely propagating muons scattered only by impurities below 3 K. Finally the data on the electric field gradient in $AlMn_{57\ ppm}$ at 40 mK show that the muon is localized at or near a single interstitial site in contrast to the idea of free propagation between the impurities.

Thus, uniform diffusive motion is not able to account for the observations. Therefore we consider trapping preceded by coherent motion, which accelerates with decreasing T and provides an increasing trapping rate with decreasing temperature. This process, again, can be described in the framework of the two-state model. Two simplifications can be made: (i) in view of the complete motional narrowing in pure Al, the differential equation (3.24) can be considered in the limit of $\tau_c \to 0$; (ii) since release processes which require activation can be neglected at low T, the limit $\tau_0 \to \infty$ can also be applied. Then the main source of depolarization comes from muons which are trapped and depolarize with the static linewidth. The fraction of trapped muons is determined by $1/\tau_1$. Contrary to depolarization during diffusion, where the depolarization is roughly proportional to τ_c (3.8), now it is roughly proportional to $1/\tau_1$; or considering (2.31), it is governed by $1/\tau_c$. With this approach the best fit for the trapping rate τ_1^{-1} yields

$$\tau_1^{-1} = 3.0(6)\ T[\mathrm{mK}]^{-0.61(4)}\ n_{Mn}\ [\mathrm{ppm}]^{0.76(6)}\ 10^5\ [\mathrm{s}^{-1}] \quad . \tag{3.28}$$

As outlined above the signs for the exponents in T and n_{Mn} are reversed to the case of τ_c^{-1} and the magnitudes agree with those of (3.27).

Depending on the magnitude of the mean free path ℓ, muon trapping by coherent diffusion may be either capture or diffusion controlled. Capture-limited processes require $\ell >> r_t$ which, following the discussion above, seems very unlikely. Still, if this process were active, the capture rate $\tau_1^{-1} = \sigma v n_t$ would be independent of temperature for deep traps (2.32). This is not observed. For structured traps temperature dependencies of σ_{abs} more complicated than $\sigma_{abs} \sim 1/v$ could also be envisaged, but they are expected to be sensitive to the type of impurity contrary to the experimental finding of an almost universal behavior.

Therefore diffusion-limited trapping preceded by coherent motion with a relatively small free path — there is no sign for a crossover to capture limited trapping —

is most likely the motional mechanism underlying the experimental findings. According to (2.31) the capture rate $1/\tau_1$ is proportional to $n_t D$. Considering the magnitude of J_{eff} ($zJ_{eff} \gg kT$), the coherent diffusion coefficient $[D_c \cong v^2\tau;\ (2.23)]$ should be governed by the thermal velocity $v \sim T^{+1/2}$ and an inverse transport scattering rate. Regarding the result for τ_1^{-1}, $\tau^{-1} \sim T^{1.6}\, c^{0.24}$ can be deduced formally.

This temperature dependence is in obvious disagreement with the prediction of KAGAN and KLINGER /2.29/ for phonon scattering (2.26) which reads $\tau^{-1} \sim T^9$. On the other hand, it is in fair agreement with a scattering rate expected from electron scattering $\tau^{-1} \sim T^2$ (Sect.2.5.2). However, neither the residual concentration dependence nor the exact temperature power are understood. Also the reason why at low T the trapping sites in AlMn are octahedral, while at higher T they are tetrahedral is not known. KEHR et al. /2.27/ argue that at low T the particle may not be able to approach the Mn impurity as closely as at higher T and therefore be trapped at sites further out not affected as much by the impurity. Such an explanation should be tested by measurements on Al doped with, e.g., Ag, which provides only short-range disturbances, and where one expects trapping to occur always in the close neighborhood of the impurity. We note that the depolarization decays observed in $AlMn_{42}$, in $AlLi_{75}$ and in $AlAg_{117}$ are about equal. This suggests trapping radii exhibiting approximately the relation $r_{Mn} : r_{Li} : r_{Ag} = 3 : 2 : 1$.

Starting from (2.31) and for the sake of simplicity again assuming a trapping radius of $r_t \cong a$, the magnitude of the coherent diffusion coefficient can be estimated to

$$D_c \cong 1.5 \cdot 10^{-7}\ T[K]^{-0.61}\ [cm^2/s] \quad . \tag{3.29}$$

Finally we ask whether similar to Nb barriers to self-trapping could also possibly account for the observed low-T structure of the depolarization rate $\sigma(T)$ which is characterized by deep minima around 3 K. In Sect.2.7 an estimate of $A = zJ_f/\hbar\omega_D$ (2.34) for Al revealed a value around one and a barrier to self-trapping cannot be discarded from the outset. Such a barrier would manifest itself in two ways in the experimental data: (i) impurities tend to reduce the barrier and small-polaron formation is facilitated in the vicinity of the impurity. With growing impurity concentration an increasing fraction of muons is expected to localize at the impurity. No temperature dependence of σ can be anticipated until the muon escapes the trap by thermal activation which occurs at higher T (3.20). (ii) Also thermal fluctuations may preform self-trapping distortions. Equation 2.35 reveals that below $\theta_D/2$ they are provided by zero-point vibrations and no temperature dependent self-trapping mechanism can be expected from thermal fluctuations at the low temperatures in question. Contrary to these predictions, the μSR results on doped Al exhibit strong temperature effects in $\sigma(T)$. There are no plateau regions in σ as found for the case of Nb (Sect.5.2.2). Thus, there is no hint of any significant delay of small-polaron formation, but the data are naturally explained by diffusion-limited trapping.

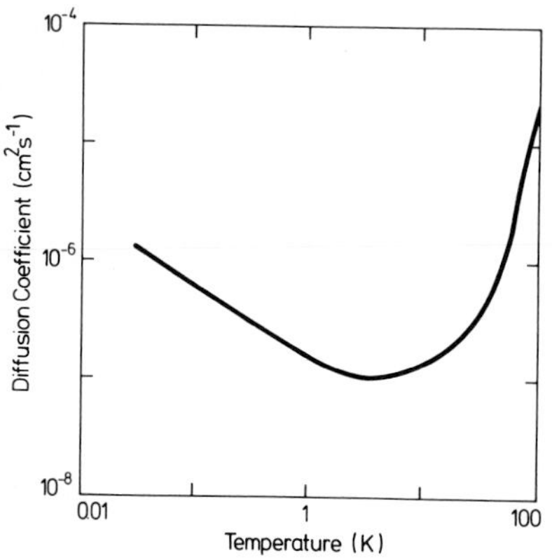

Fig.3.25. Muon diffusion coefficient in Al derived from trapping data at various impurities and temperatures

Muon diffusion in Al — summary

Figure 3.25 summarizes the results for the muon diffusion coefficient in aluminium. They represent the most extended solid-state diffusion study ranging over nearly 4 orders of magnitude in temperature. Commencing at the high-temperature side, first the diffusion drops rapidly with falling T as a consequence of the activated behavior above 60 K. Around 50 K the decrease is slowed down and the diffusion coefficient is nearly proportional to the temperature. Between 2 and 3 K D exhibits a minimum and rises again toward lower temperatures — a unique feature of coherent diffusion. Down to 30 mK the temperature dependence follows over nearly 2 order of magnitude a simple power law in T showing no signs of any further crossover at lowest T. The essential results of these investigations are (i) that clear evidence for both incoherent hopping and coherent motion of muons in Al has been found, and (ii) that small-polaron theory appears to experience great difficulties in describing the results in an adequate form. Contrary to the experimental results, small-polaron theory in its conventional form predicts strong temperature dependencies of coherent and incoherent diffusion and can account only for the muon results above 60 K. The linear temperature dependence found between 2 and 20 K might be explained by one-phonon assisted jumps between O and T sites in Al, though the quantitative estimates are grossly wrong. In the coherent range below 3 K phonon interaction does not explain the weak temperature dependencies, and electron scattering has been invoked as a possible mechanism.

4. Hydrogen Diffusion and Trapping in Metals

Contrary to the exploration of muon diffusion, the study of hydrogen in metals has a long history and dates back to 1866, when GRAHAM observed that Pd can take up H in large quantities /4.1/. Since then an enormous amount of work has been dedicated to this subject including both applied as well as fundamental research /4.2/. In view of this large body of knowledge, this review cannot attempt to present a comprehensive survey on hydrogen diffusion studies, but rather will emphasize representative results on selected bcc and fcc metals including some data on intermetallic compounds like $LaNi_5$ and $Ti_{1.2}Mn_{1.8}$, in order to elucidate the present status of hydrogen diffusion studies from a basic research point of view.

At the beginning we concentrate on the temperature and isotope dependence of the H diffusion coefficient in various materials at dilute H concentrations in order to learn about the nature of the elementary diffusive step. The experimental findings are discussed in terms of small-polaron theory as outlined in the second chapter.

The geometric details of the diffusional process can uniquely be accessed by quasielastic neutron scattering which reveals the microscopic features of the diffusional process in space and time on an atomic scale. After outlining the method on the example of hydrogen diffusion in dilute α-PdH_x, we turn to the case of the bcc metals. Special emphasis is imposed on the anomalous H behavior at higher temperatures, where correlated jump processes are significant.

The influence of substitutional and interstitial impurities on hydrogen diffusion properties is surveyed thereafter. The trapping capability of point defects has been concluded firstly from increasing H solubility with increasing amount of impurities. Resistivity, internal friction and Gorsky-effect experiments reveal further evidence for the trapping process. Again, quasielastic neutron scattering gives insight into the microscopic details of the diffusion process in the presence of trapping impurities. Results on NbH_x doped with nitrogen impurities are interpreted in terms of the two-state model outlined in Sect.2.6. The effect of substitutional impurities is demonstrated with the example of V impurities in Nb.

In concentrated systems, where H-H correlations and blocking are important, H diffusion is not well investigated on a microscopic level. In monoatomic compounds some results on Pd are available. The main emphasis, however, appears to concern possible hydrogen storage materials. Recent experiments on $T_{1.2}Mn_{1.8}H_3$ or $LaNi_5H_6$ showed a rather complicated diffusional process characterized by H sites exhibiting different energy levels and local jumps much faster than the transport jump rate. An interpretation of these results in terms of a three-state model similar to the two-state model applied to diffusion in the presence of traps will be discussed.

The lattice dynamics of metal-hydrogen systems and their conjunction with the jump process are important for an understanding of the fast diffusion mechanism. Information on the H potential, knowledge of which is essential for a quantitative calculation of H diffusion properties, can be obtained from the local H vibrations. Recent inelastic neutron scattering experiments on Nb and Ta yielded results on its anharmonicity and isotope dependence. The implications of the observed deep potential wells on the possible diffusion mechanisms and the importance of hydrogen motions in phase with the host lattice vibrations for lattice-activated diffusion processes are discussed.

Finally the dynamical behavior of hydrogen in a trapped state is examined. Internal friction measurements on H trapped at O impurities in Nb exhibit a complex relaxational behavior. Its temperature dependence appears to be in agreement with small-polaron theory. Measurements of the local H vibrations of protons trapped at substitutional and interstitial impurities in Nb revealed the astonishing result of vibrational levels very similar to those in the dilute α phase. The symmetry of the trapping sites closely resembles that of tetrahedral sites in an only weakly disturbed surrounding. Specific heat experiments on the NbN_xH_y system show a Schottky-type anomaly at about 1 K with a strong isotope effect. Recent neutron scattering experiments visualized directly the underlying H tunneling process. At higher temperatures quasielastic neutron scattering revealed a local jump process of the trapped protons. Its q dependence points to the direction of nearest-neighbor jumps in a double well potential.

4.1 Temperature and Isotope Dependence of the Hydrogen Diffusion Coefficient in Selected bcc and fcc Metals

The experimental methods for determination of the hydrogen diffusion coefficient can be divided into two groups depending on whether the coefficient is obtained from the relaxation of a nonequilibrium distribution or is investigated under equilibrium conditions. The first group encloses the commonly applied macroscopic methods like Gorsky effect, experiments of permeation, resistivity relaxation, heat of transport, etc. These methods have been reviewed elsewhere /4.3,4/ and will not be described here. The second group comprises microscopic experiments like nuclear magnetic resonance, Mössbauer effect and quasielastic neutron scattering (QNS). The large number of experimental results obtained by the various methods have been compiled recently /4.5/ and are presented in Table 4.1 for our examples. In all cases, the data are represented in the form of an Arrhenius relation: $D = D_0\exp(-E_a/kT)$ where E_a is the activation energy for the diffusive process. In the case of Pd and

Table 4.1. Activaction energies E_a and preexponential factors D_0 of the Arrhenius relation for the H(D,T) diffusion coefficients for dilute concentrations in various metals

System	E_a [meV]	$D_0 \cdot 10^4$ [cm^2/s]	System	E_a [meV]	$D_0 \cdot 10^3$ [cm^2/s]
Nb-H					
T > 270 K /4.5/	106	5.0	Pd-H /4.5/	230	2.9
T < 250 K /4.5/	68	0.9	Pd-D /4.6/	206	1.7
Nb-D /4.5/	127	5.2	Pd-T /4.7/	186	0.75
Nb-T /4.5/	135	4.5			
			Cu-H /3.38/	403	11.3
Ta-H			Cu-D /3.38/	382	7.3
T > 270 K /4.5/	140	4.4	Cu-T /3.38/	378	6.1
T < 200 K /4.5/	40	0.02			
Ta-D /4.5/	160	4.6	Ni-H /3.38/	409	7.0
			Ni-D /3.38/	401	5.3
V-H /4.5/	45	3.1	Ni-T /3.38/	395	4.3
V-D /4.5/	73	3.8			

Ni, the various methods reveal remarkably consistent values for the diffusion constant. This can be attributed to the relatively well-defined surface conditions which lead to reliable results also for the different permeation techniques.

For the other materials, a good consistency of the data exists only if the results of surface independent methods are compared. For the case of H in Nb, Fig.4.1 presents selected data obtained by different methods /4.8-13/. Recent permeation results on Pd-plated Nb samples are included /4.13/. Figure 4.2 /4.14/ summarizes H(D) diffusion results for the bcc metals Nb, V and Ta and for fcc Pd. The most striking feature is the pronounced decrease of the activation energies for H diffusion at low temperatures as observed for H in Nb and Ta, but not for D and T in the investigated temperature range. In V, where the activation energy at high temperatures has the lowest value of all three bcc metals, no change of slope has been observed. Recently indications of such a change in the activation energy for H diffusion has also been reported for fcc β-$PdH_{0.71}$ /4.15/, where a change of E_a from a high-T (T > 250 K) value of 210 meV to a low T (T < 200 K) value of 130 meV was observed.

Concerning the isotope dependence of the diffusion coefficient which primarily gives hints to the nature of the diffusion mechanism, we have the following situation for the 3 bcc metals: H always diffuses faster than D and T in the whole in-

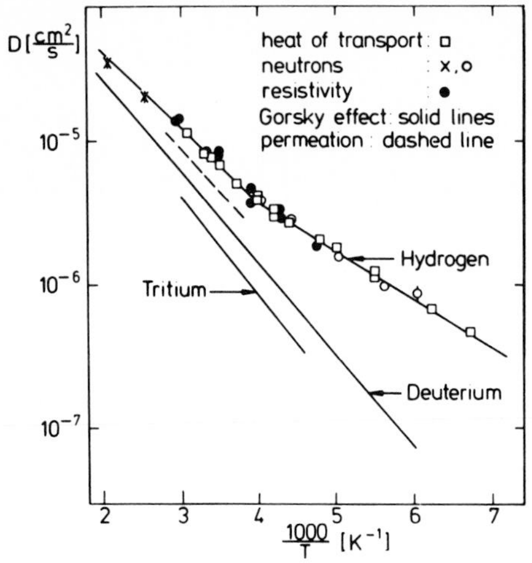

Fig.4.1. Temperature dependence of the diffusion coefficients of H, D and T in Nb: Heat of transport /4.8/, neutrons /4.9,10/, resistivity /4.11/, Gorsky effect /4.12/, permeation /4.13/

▼ Fig.4.2. H(D) diffusion coefficients from Gorsky-effect measurements in V, Nb, Ta and Pd

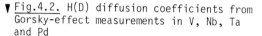

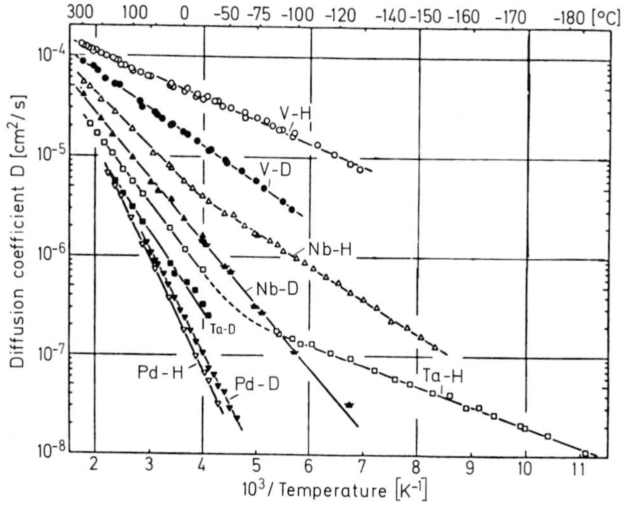

vestigated temperature range. The preexponential factors D_0 in the high-temperature range are nearly independent of the isotope and if one corrects for the lattice constant a the preexponential $1/\tau_0 = D_0 a^2/48$ is also independent of the material. For the activation energies $E_H < E_D < E_T$ holds. The ratio of the diffusion coefficients of hydrogen and deuterium are in general far away from the classical prediction: $D_H/D_D = \sqrt{m_D/m_H}$.

As an example of the temperature and isotope dependence of H diffusion in fcc metals, Fig.4.2 presents the results for Pd. In contrast to the behavior in the bcc metals, where the jump rates increase with decreasing mass of the isotope, here the heavier isotopes diffuse faster than the lighter ones /4.5-7,16/. An extrapolation of the high-temperature results of KATZ et al. /3.38/ for Ni and Cu would result in a similar reversed isotope effect for the diffusion coefficients at lower temperatures yielding $D_H < D_D < D_T$. For all three fcc metals the preexponentials D_0 for H and D behave close to $\sqrt{m_D/m_H}$. This holds also for T in the case of Ni and Cu; i.e., $D_{0H}:D_{0D}:D_{0T} = 1:1/\sqrt{2}:1/\sqrt{3}$, while for Pd D_{0T} appears to be significantly smaller. Contrary to the bcc metals, the activation energies decrease in going from H to T, causing the reversed isotope effect at low temperatures.

As can be seen from Fig.4.2 and Table 4.1 the diffusion coefficient in fcc metals are generally smaller than those in bcc materials. This structural influence has been demonstrated using the example of PdCu, which exists in a bcc as well as in a metastable fcc phase /4.14/. In Fig.4.3 the results for the H diffusion coefficient in both phases are shown together with the diffusion coefficients in pure Cu and Pd. A dramatic increase of the diffusion coefficient by nearly four orders of magnitude at room temperature becomes evident in going from the fcc to the bcc phase. It results from a strong decrease of the activation energy by a factor of 10 in the bcc phase: $E_a^{bcc} = 35 \pm 3$ meV and $E_a^{fcc} = 330 \pm 15$ meV.

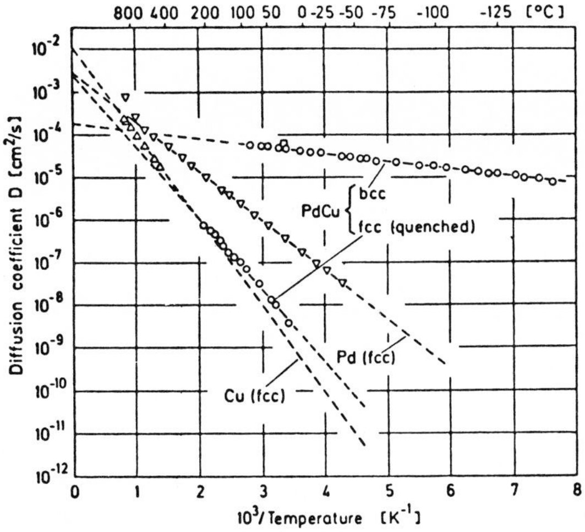

Fig.4.3. H diffusion coefficients in $Pd_{0.43}Cu_{0.57}$ (fcc and bcc phase) and in Pd and Cu

It is evident that the temperature and isotope dependencies of H diffusion in bcc and fcc metals cannot be understood in terms of classical diffusion theory predicting isotope-independent activation energies and prefactors scaling with $1/\sqrt{m}$ (2.4). In the following we investigate to what extend the quantum-mechanical jump theories as outlined in Chap.2 are able to account for the experimental data.

For the bcc metals, which are characterized by diffusion anomalies at low temperatures and isotope-independent prefactors as well as isotope-dependent activation energies at high T, the Flynn-Stoneham theory in its simplest form (2.2) fails to explain the results even qualitatively. Equation (2.6) predicts a prefactor proportional to J^2 very sensitive to the respective isotope, while the experimental prefactors do not depend on the isotope masses. The observed isotope effect in the activation energy $E_H < E_D$ again is in conflict with theory. For H and D in Nb the double force tensors are equal /4.17/ and no isotope effect is expected. With respect to the double force tensors in Ta ($P_D < P_H$) /4.17/ $E_D < E_H$ should have been observed. Thus small-polaron theory within the Condon approximation does not describe the results for Nb and Ta above 250 K. An extension of the theory /4.18/ using the concept of extended hydrogen wave functions in order to account for the discrepancies disagrees with proton form factor measurements by neutron scattering /4.19/.

Recently, the consideration of tunneling matrix elements influenced by phonon fluctuations, which significantly alter the potential barrier, shed a new light on the situation (Sect.2.3). As shown in Fig.2.4, EMIN et al.'s calculations /2.19/ on a one-dimensional model simulating Nb reproduce well the observed diffusional behavior. Not only the high-T results but also the change of the activation energy for H which is ascribed to the influence of nonadiabatic transfers are predicted correctly. Similarly, the investigations of TEICHLER, who treated the influence of phonon fluctuations on J more rigorously, point to the same direction /2.20/ (App.1). For strong correlations between the fluctuations of the interstitial levels and the barrier height a sharp transition between nonactivated behavior at low T and activated behavior at higher T is expected which could explain the observed changes of the activation energies.

In fcc metals H diffusion shows a more classical behavior, though the inverse isotope effect in the activation energies cannot be understood from classical theory. Early interpretations were given in the framework of semiclassical theories which consider the discrete structure of the H vibrational levels in the ground state and the saddle-point configuration. The diffusion data of H, D and T in Ni and Cu, e.g., were so analyzed, and the inverse isotope effect exhibited by the activation energies was related to higher frequencies at the saddle point rather than in the ground state /2.38/. However, unrealistically high local mode frequencies resulted (Cu: 300 meV in the ground state, 470 meV at the saddle point). Placing more emphasis on the structural properties, BOHMHOLDT and WICKE /4.20/ proposed an O-T-O path for H in Pd which was used by SICKING /4.21/ in order to explain the anomalous isotope dependence. According to him it originates from the discrete spacing of the vibrational levels, where more or less by accident the first vibrational level allowing tunneling to a neighboring site happens to be lower for deuterium than for hydrogen. Thereby he neglected small-polaron effects and the influence of phonon fluctuations.

Recently TEICHLER reinvestigated the problem /4.22/, applying small-polaron theory beyond the Condon approximation as already proposed by FLYNN and STONEHAM /2.4/ and outlined in Sect.2.3. In order to reach quantitative results, he used his microscopic model for a H potential in Cu derived from pseudopotential calculations /2.17/. This potential was able to describe μ^+ diffusion properties /3.5/ and the forces exerted by the muon on the Cu atoms (3.4) reasonably well. Figure 4.4 represents the potential along the O-T-O path for the unrelaxed lattice. The bottom of the T well lies 640 meV above that of the O site. The saddle point is 850 meV above the O minimum. Within the O potential well several vibrational states exist, while the T site contains only one excited state. The most significant feature is the large difference of 300 meV between the relaxation energies ΔU (2.1) at tetrahedral and octahedral sites. Thus the proton may significantly reduce the energy barrier between O sites by occupying intermediately an excited state at the T site. The lattice relaxation associated with the intermediate T site "opens the door" for the diffusive jump (Sect.2.3).

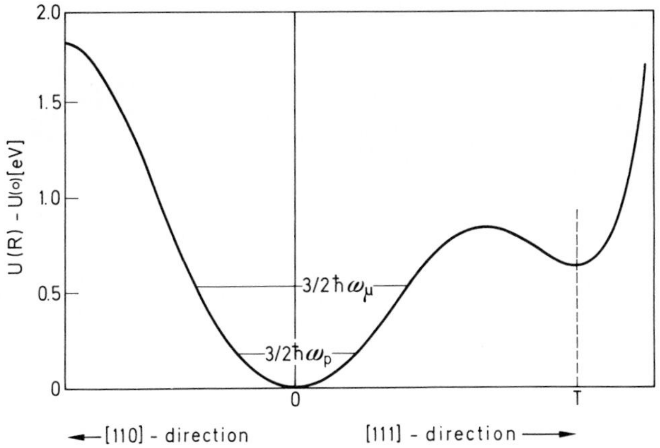

Fig.4.4. H potential in Cu along an O-T path according to TEICHLER's pseudopotential calculations /2.17/

Starting from (A1.1) and reducing the problem to one dimension along the O-T-O path the H diffusion coefficient was evaluated. The calculations reveal that above 150 K lattice-activated processes dominate. Table 4.2 compares the theoretical activation energies with the experimental values of KATZ et al. /3.38/. Though the absolute numbers disagree, the theory reproduces well the observed trend of a decreasing activation energy with increasing mass. For the more complicated transition metals Ni and Pd no theoretical calculations are available so far.

Table 4.2. Experimental activation energies for H(D,T) diffusion in Cu /3.38/ compared to theoretical values according to TEICHLER /4.22/

Isotope	E_a (theoretical) [meV]	E_a (experimental) [meV]
H	282	403
D	270	382
T	264	378

4.2 Quasielastic Incoherent Neutron Scattering Studies of H Diffusion

4.2.1 Outline of the Method

In order to study the diffusive motion of a particle in all its details a probe is needed which offers a space resolution of atomic distances and at the same time provides a time resolution in the order of inverse diffusional jump rates. Quasielastic neutron scattering (QNS) fulfills both requirements and is the only method which allows the simultaneous investigation of the time and space development of the diffusive process on the level of elementary jumps. In a QNS experiment the neutron spectrum centered around the energy transfer zero is analyzed as a function of energy transfer $\hbar\omega$ and scattering angle 2ϑ. Its broadening in energy bears information on the H jump rate as well as on the diffusion coefficient. The dependence on the scattering angle and the details of the lineshape contain information on the geometrical way in which H diffusion proceeds, including the information on the occupied interstitial site /4.23/.

As a consequence of the spin-dependent scattering lengths for the neutron-proton interaction and the random orientation of the proton spins at all temperatures of interest, H scatters predominant incoherently – neutrons scattered from different protons have random phase relations and do not interfere. Thus, a QNS experiment on H reveals information on the motional behavior of single protons. Then following the concept of VAN HOVE /4.24/ the double differential neutron cross section $\partial^2\sigma/\partial\omega\partial\Omega$ is proportional to the Fourier transform of the proton self-correlation function $G_s(\underline{r},t)$:

$$\frac{\partial^2\sigma}{\partial\omega\partial\Omega} = \frac{\sigma_{inc}^H}{4\pi}\frac{k_f}{k_i}\frac{1}{2\pi}\int_{-\infty}^{+\infty}dt\int_{-\infty}^{+\infty}d^3\underline{r}G_s(\underline{r},t)e^{i(\underline{q}\underline{r}-\omega t)} \tag{4.1}$$

$$= \frac{\sigma_{inc}^H}{4\pi}\frac{k_f}{k_i}S_{inc}(\underline{q},\omega) \quad ,$$

where $\hbar \underline{q} = \hbar(\underline{k}_f - \underline{k}_i)$ is the momentum transfer which for elastic scattering is related to the scattering angle by $q = (4\pi/\lambda)\sin\vartheta$ (λ = neutron wavelength), $S_{inc}(q,\omega)$ is the so-called incoherent scattering law, and σ_{inc}^{H} denotes the incoherent H cross section. It is worthwile to note that σ_{inc}^{H} surpasses the cross section of most other atoms by one or two orders of magnitude, so that scattering experiments can be performed even at H concentrations of some tenth of a percent in favorable cases.

In the following we treat the self-correlation function in the classical limit which is justified for diffusional motion if no phase relation between the hydrogen wave functions at different sites (or intermediate states) exist. In this case $G_s(\underline{r},t)$ can be interpreted as the conditional probability to find a proton at time t at a site $\underline{r}$, if it has been at $\underline{r} = 0$ for $t = 0$. The results of QNS experiments are commonly interpreted in more or less generalized versions of the Chudley-Elliott model (CE) /4.25/ which introduces the following basic assumptions:

(i) the jump time from site to site is small compared to the mean rest time of hydrogen on its interstitial site; (ii) no correlations between diffusive motion and thermal vibrations of the particle at its equilibrium site exist.

For the sake of clarity we first assume a Bravais H sublattice. Then, in the framework of the CE model and for infinite dilution the self-correlation function for the diffusive motions $G_s^D(\underline{r},t)$ obeys the master equation

$$\frac{\partial}{\partial t} G_s^D(\underline{r},t) = \frac{1}{z} \sum_{i=1}^{z} \frac{1}{\tau_i} \left[G_s^D(\underline{r}+\underline{s}_i,t) - G_s^D(\underline{r},t) \right] \quad , \tag{4.2}$$

where $\underline{s}_i$ denote the jump vectors to the accessible neighboring sites, z is the number of these sites and $1/\tau_i$ is the corresponding jump rate. Equation (4.2) can be integrated by Fourier transformation and yields for nearest-neighbor jumps ($1/\tau_i = 1/\tau$)

$$\frac{\partial}{\partial t} G_s^D(\underline{q},t) = \frac{-1}{z\tau} \sum_{i=1}^{z} (1 - e^{i\underline{q}\underline{s}_i}) \, G_s^D(\underline{q},t) \tag{4.3}$$

$$= -\frac{1}{\tau} f(\underline{q}) \, G_s^D(\underline{q},t) \quad . \tag{4.4}$$ [4]

Applying the appropriate initial condition $G_s^D(\underline{r},0) = \delta(r)$ we get

$$G_s^D(\underline{q},t) = \exp[-(t/\tau) f(\underline{q})] \quad . \tag{4.5}$$

[4] f(q) can also be considered as Fourier transform of the spatial H distribution after one jump.

Finally, Fourier transformation with respect to time yields the scattering function

$$S_{inc}^{D}(\underline{q},\omega) = \frac{1}{\pi}\frac{f(\underline{q})/\tau}{[f(\underline{q})/\tau]^2+\omega^2} \quad . \tag{4.6}$$

As a consequence of the second assumption of the CE model the vibrational motions enter the scattering function in form of a Debye-Waller factor

$$\left(\frac{\partial\sigma}{\partial\Omega}\right)_{el}^{vib} \sim \exp\left[-\frac{1}{3}q^2\langle u^2\rangle\right] \quad , \tag{4.7}$$

where $\langle u^2\rangle$ is the mean square displacement of the proton at its site.

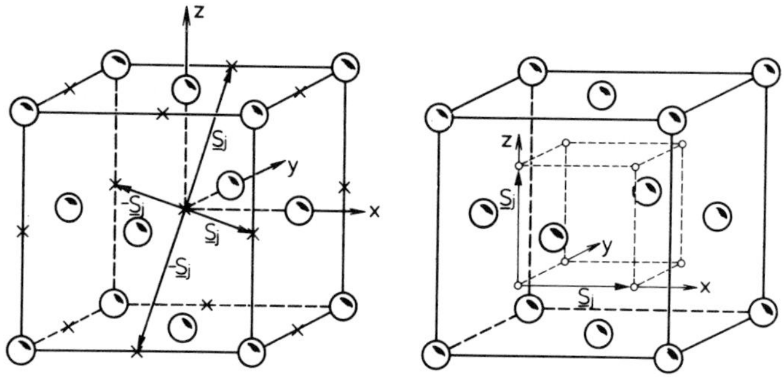

Fig.4.5. Possible H interstitial sites in Pd: (◔) Pd atoms, (x) octahedral sites, (o) tetrahedral sites, ($\underline{S}_j$) H jump vectors

Treating the diffusion of hydrogen in Pd as an example, we outline briefly the kind of information which can be obtained from an investigation of the $\underline{q}$ and ω dependence of the quasielastic scattering law. Figure 4.5 displays the two possible interstitial sites for hydrogen in fcc Pd together with the jump vectors to nearest-neighbor sites. Assuming nearest-neighbor jumps only, the functions $f(\underline{q})$ are readily calculated for octahedral as well as tetrahedral site occupation and yield

$$\begin{aligned} f_0(\underline{q}) = \frac{1}{6}\Big[6 &- \cos\tfrac{a}{2}(q_x+q_y) - \cos\tfrac{a}{2}(q_x-q_y) \\ &- \cos\tfrac{a}{2}(q_x+q_z) - \cos\tfrac{a}{2}(q_x-q_z) \\ &- \cos\tfrac{a}{2}(q_y+q_z) - \cos\tfrac{a}{2}(q_y-q_z)\Big] \end{aligned} \tag{4.8}$$

$$f_t(\underline{q}) = \frac{1}{3}\left[3-\cos(\tfrac{a}{2}q_x) - \cos(\tfrac{a}{2}q_y) - \cos(\tfrac{a}{2}q_z)\right] \tag{4.9}$$

where a is the lattice constant. We note that in order to reveal equal diffusion coefficients, the jump rate between T sites $1/\tau_t$ has to be twice as large as the rate between O sites $1/\tau_0$. For small momentum transfers the quasielastic linewidth $\Gamma(\underline{q}) = (1/\tau)f(\underline{q})$ becomes independent of the particular jump model

$$\Gamma(\underline{q}) = Dq^2 \qquad (4.10)$$

and yields the *macroscopic* diffusion coefficient measured over *microscopic* distances. The q dependence of the linewidth at larger q is sensitive to the geometrical details of the jump mechanism. Figure 4.6 displays the theoretical $\underline{q}$ dependence of linewidths for both models in (100) and (110) directions, and large differences are evident. Figure 4.6 also presents experimental data obtained by ROWE et al. /4.26/ which are clearly in favor of H jumps between nearest-neighbor octahedral sites; an intermediate occupation of a T site as proposed by TEICHLER /4.22/ (Sect.4.1) does not show itself in neutron scattering as long as $\tau_0 \gg \tau_t$.

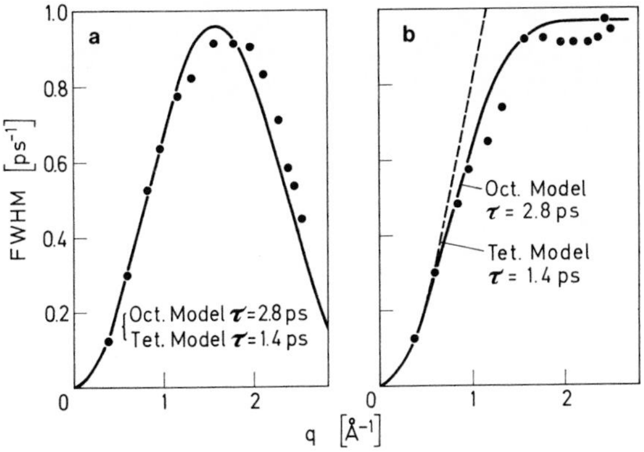

Fig.4.6a,b. Quasielastic linewidths in PdH_x for (100) (a) and (110) (b) directions

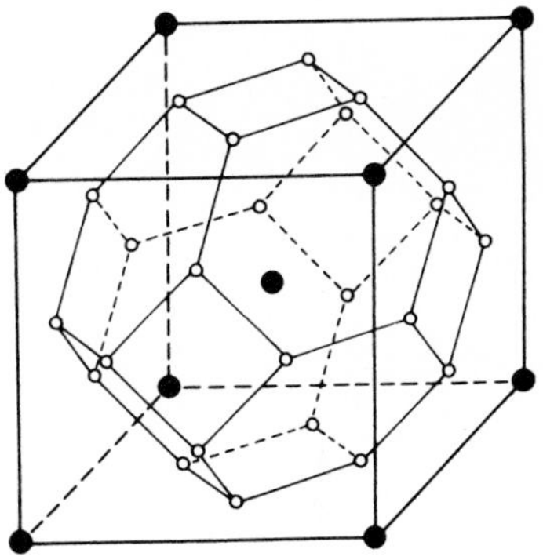

Fig.4.7. Sublattice formed by the tetrahedral sites in bcc structures

4.2.2 Correlated Jumps and Hydrogen Diffusion in bcc Metals

In bcc metals the interstitial sites form complicated non-Bravais lattices, as shown for tetrahedral sites in Fig.4.7. There are 6 nonequivalent sites per unit cell and consequently the rate equation (4.2) has to be replaced by a system of coupled differential equations

$$\frac{\partial}{\partial t} G_i^D(\underline{r},t) = \frac{1}{z\tau} \sum_{j,k} \{G_j^D(\underline{r}+\underline{s}_{ijk},t) - G_i^D(\underline{r},t)\} \quad , \qquad (4.11)$$

where for the sake of clarity we assume next-neighbor jumps from the outset. Here G_i^D is the probability to find a particle on an interstitial site with symmetry character i, and $\underline{s}_{ijk}$ denotes the jump vector from site i to site k with symmetry j. The sum is carried out over all neighboring sites. Again Fourier transformation has to be applied and reveals

$$\frac{\partial}{\partial t} G_i^D(\underline{q},t) = \frac{1}{z\tau} \sum_j \left[G_j(\underline{q},t) (\sum_k e^{i\underline{q}\underline{s}_{ijk}} - G_i(\underline{q},t)) \right] \quad . \tag{4.12}$$

In matrix notation (4.12) reads

$$\frac{\partial}{\partial t} \underline{G}^D = \frac{1}{\tau} \underline{\underline{D}}\, \underline{G}^D \quad , \tag{4.13}$$

where D is 6×6 matrix with the elements

$$D_{ij} = \frac{1}{z} \sum_k e^{i\underline{q}\underline{s}_{ijk}} - \delta_{ij} \tag{4.14}$$

A general solution for $G_i^D(\underline{q},t)$ is given by

$$G_i^D(q,t) = \sum_{k=1}^{6} \alpha_i^k e^{-(t/\tau) f_k(\underline{q})} \quad , \tag{4.15}$$

$f_k(\underline{q})$ being the eigenvalues and α_i^k the eigenvectors of the jump matrix $\underline{\underline{D}}$. Introducing $G_{ij}^D(q,t)$ by the appropriate initial condition

$$G_{ij}^D(q,0) = \delta_{ij} \tag{4.16}$$

one gets

$$G_{ij}^D(q,t) = \sum_{k=1}^{6} \alpha_i^k \alpha_j^k e^{-(t/\tau) f_k(\underline{q})} \quad . \tag{4.17}$$

The self-correlation function is obtained from an average over the initial states and a sum over the final states

$$G^D(\underline{q},t) = \frac{1}{6} \sum_{i,j,k}^{6} \alpha_i^k \alpha_j^k e^{-(t/\tau) f_k(\underline{q})} \quad . \tag{4.18}$$

After Fourier transformation with respect to time the incoherent scattering law finally becomes

$$S_{inc}^D(\underline{q},\omega) = \frac{1}{6\pi} \sum_{k=1}^{6} (\sum_i \alpha_i^k)^2 \frac{f_k(\underline{q})/\tau}{[f_k(\underline{q})/\tau]^2 + \omega^2} \quad . \tag{4.19}$$

Explicit results were first given by BLAESSER and PERETTI /4.27/ and can easily be extended along the same scheme to more than next-nearest-neighbor jumps or other non-Bravais geometries.

Besides the complications arising from the non-Bravais H sublattice in bcc metals, there is experimental evidence that the diffusion process itself is not as simple as for fcc metals. Already earlier results on V /4.28/, Ta /4.29/, and Nb /4.30/ showed systematic deviations from the predictions of a simple nearest-neighbor jump model. Also anomalies in the intensity of the quasielastic line have been reported /4.31,28/. These anomalies have been attributed to the occurrence of further-neighbor jumps and/or to jumps between different types of interstitial sites /4.28,29/ and to the influence of a finite time of flight /4.32/, but no conclusive picture has been reached.

Recently LOTTNER et al. /4.19,33,34/ have reexamined the problem for Ta, Nb, and V. Quasielastic neutron scattering experiments were performed at $NbH_{0.02}$, $TaH_{0.13}$ and $VH_{0.07}$ single crystals at temperatures between 290 and 760 K for q values between 0.3 and 2.5 $Å^{-1}$. The data were analyzed in terms of 4 different models always assuming that H jumps occur between tetrahedral sites.

Model (1): Hydrogen jumps occur only between nearest-neighbor sites as outlined above. The jump rate is $1/\tau_1$.

Model (2): In extension of Model (1) jumps to second-nearest neighbors are also included. They occur in (100) direction across the center of the cube face with a jump rate $1/\tau_2$.

Model (3): Assumes that in addition to nearest-neighbor jumps correlated double jumps are also possible. Thereby the rest time at the intermediate site is presumed to be negligible compared to the interval between successive double jumps. Consequently, the double jump can be represented by a single jump rate $1/\tau_2$ and jump vectors leading to all topologically second-nearest neighbors.

Model (4): Generalizes Model (3) and considers explicitly the rest time at the intermediate sites. This is done in the framework of a two-state model similar to that used for the description of diffusion in the presence of trapping impurities. Here it considers the H alternatively in a mobile "state" (lifetime τ_e) where it can perform repeated jumps to nearest neighbors with a jump rate $1/\tau_1$, and in an immobile self-trapped state (lifetime τ_t). The exchange between both "states" is described by transition rates given by the inverse lifetimes.

The jump matrices corresponding to the different models were evaluated in a similar manner as described above and are 6×6 matrices for Models (1) to (3) and a 12×12 matrix for Model (4) /4.33/. Model (1) includes 2 independent parameters, the jump rate $1/\tau_1$ and the quasielastic intensity, Models (2) and (3) have 3 parameters, and finally (4) is a four-parameter model. Figure 4.8 shows the results of

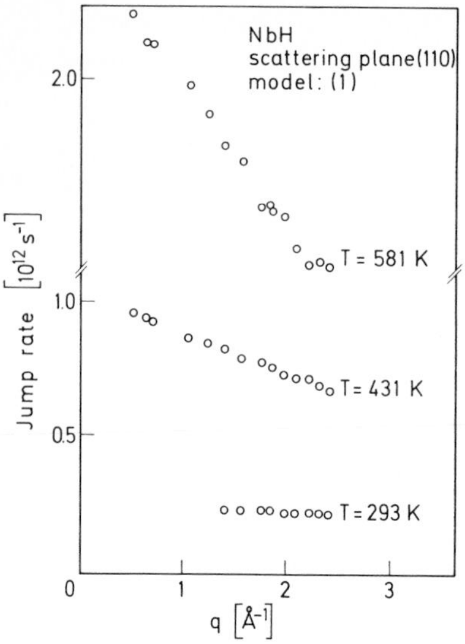

Fig.4.8. Jump rate $1/\tau_1$ of Model (1) as a function of q determined from measurements at 293 K, 431 K and 581 K

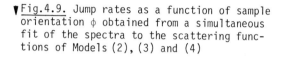

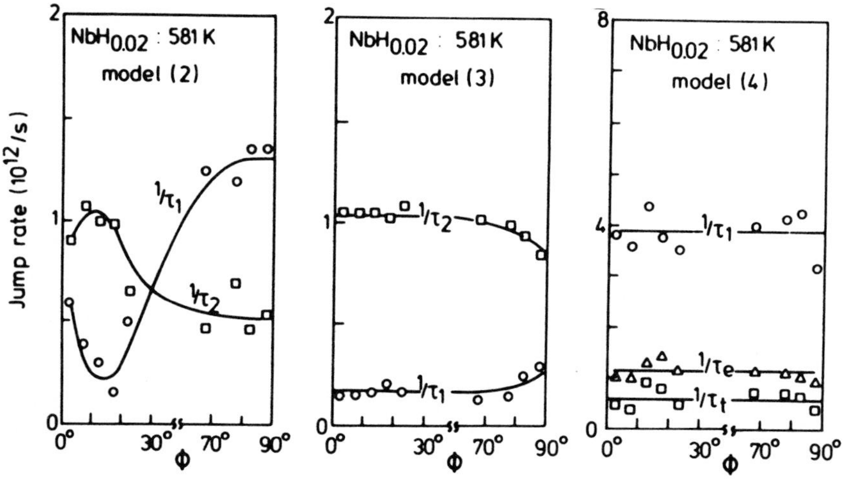

Fig.4.9. Jump rates as a function of sample orientation ϕ obtained from a simultaneous fit of the spectra to the scattering functions of Models (2), (3) and (4)

a data analysis applying Model (1). While at room temperature all spectra yield the same rate $1/\tau_1$, at higher temperatures severe deviations appear. The decrease of $1/\tau_1$ with increasing q can be understood quantitatively as an increase of the effective jump length. The diffusion coefficient given by the linewidth at small q is too large and does not correlate with the jump rate observed at large q, if it is interpreted as a mere nearest-neighbor jump. Figure 4.9 presents results obtained by an analysis of the high-temperature data with Models (2) to (4). The jump rates are calculated by a simultaneous fit of the spectra measured at different q values for one crystal orientation ϕ and are plotted vs ϕ. For a consistent description of the data, the obtained jump rates should not depend on the crystal direction.

Figure 4.9 makes it clear that the extension of Model (1) to jumps to next-nearest neighbors does not solve the problem: in order to yield orientation independent values of $1/\tau_1$ and $1/\tau_2$, the experimental spectra would have to show a much larger anisotropy than they actually exhibit. The assumption of correlated jumps on the other hand leads to a satisfactory agreement between theory and experiment. Judging and quality of the fits and taking into account that Model (4) contains an additional parameter, Models (3) and (4) can be considered to fit equally well. However, the large ratio of $\tau_1/\tau_2 = 5.5$ found at 581 K makes it physically more reasonable to assume that once the particle is in a mobile mode then more than double jumps should be possible, which is the content of Model (4). Similarly, also for H in Ta and V, the simple-jump Model (1) fails to explain the data at elevated temperatures, whereas Model (3) seems to lead to an adequate description of the experimental results also for Ta and V. Model (4) has not been used for the data analysis for Ta and V. However, again the large ratio τ_1/τ_2 found for the contribution of double jumps makes the application of Model (4) desirable.

The authors do not specify the physical origin of the occurrence of correlated jumps. Several possibilities appear plausible: (i) At high jump rates the time τ_r which the lattice needs to dissipate the energy necessary for a jump may come into the range of the jump time τ. For this case, EMIN has pointed out /4.35,36/ that successive jumps are possible, which in the limit of $\tau \ll \tau_r$ require only one third of the activation energy needed for a jump in the relaxed lattice. Such behavior is suitably described by Model (4). (ii) Lattice activation may lead to correlated jumps: certain phonons or combination of phonons may open the door for more than one jump at a time. Such a process could be adequately described in a "forward jump model", where the direction of the second jump is predetermined by the first one. The authors explicitly concede that such a model would also fit their data. For a final distinction between these two possibilities (and maybe even others) quantitative calculations are desirable.

The q dependence of the quasielastic intensity evaluated from this experiment follows a normal Debye-Waller factor (DWF) behavior with mean-square amplitudes for the H motion of $\langle u^2\rangle/3 = 0.02 - 0.04$ Å^2 which is near to the value expected from harmonic calculations /4.34/. For V an anomalous decrease of intensity has been observed at larger q values and high temperature (T = 763 K) which was attributed to nonnegligible jump times /4.34/. The normal DWF evaluated for Nb and Ta even at high temperatures is in contrast to earlier reports of anomalous behavior, the latter resulting from errors in the integration of the quasielastic spectrum neglecting wing contributions. The observation of a normal DWF shows that the proton is well localized at its interstitial site in contradiction to earlier speculations about an extended proton wave function /4.18/.

4.2.3 Hydrogen Diffusion in the Presence of Trapping Impurities

The physical properties of H dissolved in bcc transition metals are strongly changed by the presence of small amounts of impurities. Vapor pressure data show deviations from Sieverts' law at low concentrations which increase by introducing oxygen impurities into Ta or due to cold working of V /4.37/. Solubility data reveal a shift of the phase boundary between the α and β phase in the presence of O, N or C impurities /4.38/. Resistivity experiments on Nb-H doped with N /4.39/ yield reduced residual H resistivity at temperatures well above the phase boundary to the β phase. Furthermore, Gorsky-effect measurements in NbN_xH_y samples /4.40/ showed a strong decrease of the H diffusion coefficient in particular at lower temperatures. Also, in the presence of O and N impurities in the Nb-H system additional relaxation peaks in internal friction experiments are observed /4.41,42/. Finally, recent inelastic neutron scattering experiments on Nb doped with the interstitial impurities O or N /4.43/ or the substitutional impurity Ti /4.44/ yielded spectra of the local vibrations which at low temperatures deviate strongly from the vibrational pattern of the ordered Nb hydrides.

All these features are naturally explained in terms of H trapping at the impurities. In their vicinity they lower the ground-state energy for the H. This gives rise to deviations from Sieverts' law at low concentrations due to an increase of the enthalpy of solution. If the binding energy at the trapping center is larger than the enthalpy of formation for an ordered hydride phase, the solubility in the dilute phase increases. Resistivity measurements suggest that in Nb<u>N</u> each N impurity is capable of keeping one H from precipitating into the hydride phase. Trapping implies the formation of impurity hydrogen pairs. They are assumed to scatter conduction electrons with a smaller probability than the two single scattering centers and they give rise to relaxation processes in internal friction experiments. Finally, H in a trapped state is expected to exhibit a vibrational behavior characteristic for the kind of trap and in general different from that of the ordered hydride.

Results on the local dynamics as revealed from internal friction measurements or inelastic neutron scattering will be treated in Sect.4.4. Here we concentrate on the effects of trapping on long-range diffusion, where the significance of trapping processes will increase with increasing ratio of binding energy and thermal energy kT.

On a microscopic level, the influence of interstitial impurities on the H diffusion process was studied by quasielastic neutron scattering on Nb-H samples doped with nitrogen impurities /2.39/. Also some preliminary neutron scattering results for the Nb/V/H system are available /4.45/. In the following we first derive the incoherent scattering function for H diffusion in the presence of traps and thereafter discuss the experimental results.

Scattering law for H diffusion in the presence of traps

For low concentrations of trapping impurities, where trapping regions and regions of undisturbed host material are present, relatively simple arguments can be given in order to explain what kind of information can be obtained from a quasielastic neutron scattering experiment: for small momentum transfers, the scattering process averages over large volumes in space [of the order $(2\pi/q)^3$]. Therefore, a long section of the diffusive path of the proton will be probed by the neutron wave packet. This path consists of periods of undisturbed diffusion as well as portions where the proton is trapped. Under these circumstances, the scattering law is expected to be a single Lorentzian whose width is given by the effective or macroscopic H diffusion coefficient D_{eff} in the system. At large q's the average occurs over short distances and the scattering law depends on the single diffusive step. In this case it contains information about the mean trapping time and on the fraction of protons being trapped.

Quantitatively, the scattering law could be calculated from a generalized version of the rate equation (4.11). The randomly distributed impurities cause the loss of the translational symmetry, and the corresponding rate equation depends explicitly on the lattice position $\underline{m}$, where $\underline{m}$ may summarize translation vector and in the case of non-Bravais lattices also the symmetry character of the site within the unit cell. We have

$$\frac{\partial}{\partial t} G_{\underline{i}}^{\underline{m}}(t) = \sum_{\underline{n}} \left(\Gamma^{\underline{m}\underline{n}} G_{\underline{i}}^{\underline{n}} - \Gamma^{\underline{n}\underline{m}} G_{\underline{i}}^{\underline{m}} \right) \quad , \tag{4.20}$$

where $G_{\underline{i}}^{\underline{m}}$ is the conditional probability to find a proton at site $\underline{m}$ at a time t if it was at a site $\underline{i}$ for t = 0, and $\Gamma^{\underline{m}\underline{n}}$ is the jump rate between the sites $\underline{m}$ and $\underline{n}$. The mean rest time for a H atom at a certain site $\underline{m}$ is then given by

$$\tau^{\underline{m}} = \{\sum_{\underline{n}} \Gamma^{\underline{n}\underline{m}}\}^{-1} \quad . \tag{4.21}$$

$G_{\underline{i}}^{\underline{m}}$ fulfills the initial condition

$$G_{\underline{i}}^{\underline{m}}(0) = \delta_{\underline{m}\underline{i}} \rho_{\underline{i}} \quad , \tag{4.22}$$

where ρ_i is the thermal occupancy of site $\underline{i}$, namely

$$\rho_{\underline{i}} = \frac{e^{-\beta E_{\underline{i}}}}{\sum_{\underline{n}} e^{-\beta E_{\underline{n}}}} \quad , \quad \beta = \frac{1}{kT} \quad . \tag{4.23}$$

The sum has to be carried out over all accessible sites of the host lattice. The ground state energy of the hydrogen is E_n which depends on the site $\underline{n}$ in the vicinity of an impurity atom. Formally, the rate equation (4.20) can be solved and the corresponding scattering law can be calculated. In general it is given by a superposition of Lorentzians similar to (4.19), where the widths λ_s are the eigenvalues of (4.20) and the weight factors $g_s(\underline{q})$ are related to the eigenvectors and the thermal occupation numbers:

$$S^D_{inc}(\underline{q},\omega) = \sum_s g_s(\underline{q}) \frac{\lambda_s/\pi}{\lambda_s^2 + \omega^2} \quad . \tag{4.24}$$

Attempts to solve (4.20) directly have been undertaken numerically /4.45/ in a lattice of 16000 Nb atoms and 128 N atoms and the results will be briefly discussed together with the experiments.

Here, we present an analytic evaluation of the scattering law in terms of the two-state model discussed in Sect.2.6 and applied to μSR results in Sect.3.6 /2.39, 4.46/. Again we characterize the complicated structure of the trapping region around the impurity by one parameter, the escape rate $1/\tau_0$. For neutron scattering purposes this should be a good approximation as long as $2\pi/q$ is smaller than the dimension of the trap r_t. The capture rate $1/\tau_1$ obeys (2.31) for diffusion-controlled trapping.[5] The calculation of the self-correlation function in terms of a random-walk model follows similar lines as the evaluation of the muon depolarization in the presence of traps (Sect.3.6.1). In addition to the muon case, where only the time dependence of P(t) was important, here we must also consider the spatial evolution of the self-correlation function.

We start from the self-correlation function for undisturbed diffusion $G_{f0}(\underline{r},t)$ as defined by (4.2) — for the sake of clarity we shall restrict ourselves to Bravais lattices, but the generalization is obvious — and the self-correlated function in the trapped state

$$G_{t0}(\underline{r},t) = \delta(\underline{r}) \quad . \tag{4.25}$$

Let $G^F_F(\underline{r},t)$ and $G^T_F(\underline{r},t)$ be the self-correlation functions for protons which have started in the free state at t = 0 and are in the free (G^F_F) or trapped state (G^T_F) irrespective of their history. Because protons other than muons start from initial conditions corresponding to thermal equilibrium, the correlation functions G^F_T and

[5] We note that for finite H concentrations trap saturation may have to be considered. Then, (2.31) has to be modified in order to account for the effectively reduced number of empty traps (see also Sect.4.2.5).

G_T^T for protons which start in a trap also have to be considered. In analogy to (3.23) we have:

$$G_F^F(\underline{r},t) = e^{-t/\tau_1} G_{f0}(\underline{r},t) + \frac{1}{\tau_0}\int_0^t dt' \int_{-\infty}^{+\infty} d^3r' \exp\left[-(t-t')/\tau_1\right] G_{f0}(\underline{r}-\underline{r}',t-t') G_F^T(\underline{r}',t')$$

$$G_F^T(\underline{r},t) = \frac{1}{\tau_1}\int_0^t dt' \int_{-\infty}^{+\infty} d^3r' \exp\left[-(t-t')/\tau_0\right] G_{t0}(\underline{r}-\underline{r}',t-t') G_F^F(\underline{r}',\underline{t}')$$

$$G_T^T(\underline{r},t) = e^{-t/\tau_0} G_{t0}(\underline{r},t) + \frac{1}{\tau_1}\int_0^t dt' \int_{-\infty}^{+\infty} d^3r' \exp\left[-(t-t')/\tau_0\right] G_{t0}(\underline{r}-\underline{r}',t-t') G_T^F(\underline{r}',t')$$

$$G_T^F(r,t) = \frac{1}{\tau_0}\int_0^t dt' \int_{-\infty}^{+\infty} d^3r' \exp\left[-(t-t')/\tau_1\right] G_{f0}(\underline{r}-\underline{r}',t-t') G_T^T(\underline{r}',t) \quad . \tag{4.26}$$

The total self-correlation function $G(\underline{r},t)$ is then given by the appropriate thermal average

$$G(\underline{r},t) = \frac{\tau_0}{\tau_1+\tau_0}\left[G_T^T(\underline{r},t) + G_T^F(\underline{r},t)\right] + \frac{\tau_1}{\tau_0+\tau_1}\left[G_F^T(\underline{r},t) + G_F^F(\underline{r},t)\right] \quad . \tag{4.27}$$

The integral equations (4.26) can be solved by Fourier transformation in space and Laplace transformation in time and yield:

$$\begin{aligned}
G_F^F(\underline{q},s) &= B(\underline{q},s) + \frac{1}{\tau_1} B(\underline{q},s)\, G_F^T(\underline{q},s) \\
G_F^T(\underline{q},s) &= \frac{1}{\tau_0} A(\underline{q},s)\, G_F^F(\underline{q},s) \\
G_T^T(\underline{q},s) &= A(\underline{q},s) + \frac{1}{\tau_0} A(\underline{q},s)\, G_T^F(\underline{q},s) \\
G_T^F(\underline{q},s) &= \frac{1}{\tau_1} B(\underline{q},s)\, G_T^T(\underline{q},s)
\end{aligned} \tag{4.28}$$

with

$$A(\underline{q},s) = \frac{\tau_0}{1+s\tau_0} \tag{4.29}$$

and with the aid of (4.5)

$$B(\underline{q},s) = \frac{\tau_1}{1+[f(\underline{q})/\tau+s]\tau_1} \quad . \tag{4.30}$$

After some algebra and using $S^D_{inc}(\underline{q},\omega) = \pi^{-1}\,\mathrm{Re}\{G(\underline{q},s=i\omega)\}$ we arrive at

$$S^D_{inc}(\underline{q},\omega) = R_1\,\frac{\Lambda_1/\pi}{\Lambda_1^2+\omega^2} + (1-R_1)\,\frac{\Lambda_2/\pi}{\Lambda_2^2+\omega^2} \tag{4.31}$$

with

$$\Lambda_{1/2} = \frac{1}{2}\left(1/\tau_0+1/\tau_1+\frac{1}{\tau}\,f(\underline{q}) \pm \sqrt{[1/\tau_0+1/\tau_1+\frac{1}{\tau}\,f(\underline{q})]^2 - 4\,\frac{f(\underline{q})}{\tau_0\tau}}\right) \tag{4.32}$$

and

$$R_1 = \frac{1}{2}+\frac{1}{2}\left[\frac{1}{\tau}\,f(q)\,\frac{\tau_1-\tau_0}{\tau_1+\tau_0} - 1/\tau_0-1/\tau_1\right] \times\sqrt{[1/\tau_0+1/\tau_1+\frac{1}{\tau}\,f(\underline{q})]^2 - 4\frac{f(\underline{q})}{\tau\tau_0}}^{\;-1} \tag{4.33}$$

In order to illustrate the typical features of the scattering law Fig.4.10 represents the behavior of its two components, namely their widths $\Lambda_{1/2}$ and weights $R_{1/2}$, as a function of Dq^2. Included is also $f(\underline{q})/\tau = \Lambda(\underline{q})$ for the undisturbed lattice. For the sake of simplicity the calculations are made for a simple cubic lattice.

For small q an expansion of (4.33) yields $R_1 = \tau_1^3 D^2 q^4/\tau_0$. Thus, in the limit $q \to 0$ the scattering law (4.31) consists of one Lorentzian. Its width can be in-

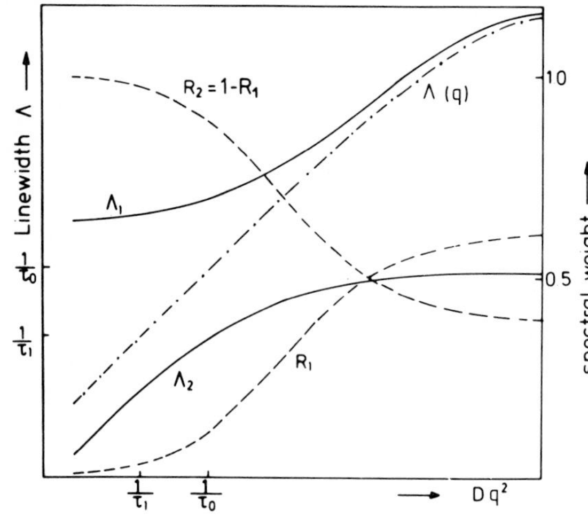

Fig.4.10. Linewidths of the two components Λ_1 and Λ_2 (———) and their spectral weights (-----) R_1 and R_2 in the scattering law for the two-state model describing diffusion in the presence of traps. Dashed dotted line: linewidth for an undisturbed lattice $\Lambda(q)$

ferred from (4.32) and is given by the effective diffusion coefficient in the host lattice with impurities

$$\Lambda_2 = D \frac{\tau_1}{\tau_1+\tau_0} q^2 = D_{eff} q^2 \tag{4.34}$$

as expected from the qualitative arguments given at the beginning on this section. For q approaching the zone boundary, the spectrum is essentially related to the single diffusion steps. Equation (4.31) leads to two Lorentzians, a narrow one whose width is given by the escape rate out of the trap $1/\tau_0$, and a broad one with a width given by the jump rates in the undisturbed lattice $1/\tau$. The weights correspond to the relative occupations of the free and the trapped state.

The results of the two-state model agree formally with an average T-matrix approximation which relates the parameters $1/\tau_0$ and $1/\tau_1$ to more microscopic quantities /2.40/. The leading term of the escape rate $1/\tau_0$ is independent of the trap concentration n_t whereas the trapping rate $1/\tau_1$ is proportional to n_t as expected from (2.31).

QNS experiments on H diffusion in the presence of traps

Quasielastic neutron scattering experiments were carried out on two Nb samples doped with N and H. The N concentrations were $n_N^1 = 0.7\%$ and $n_N^2 = 0.4\%$. The corresponding H concentrations were $n_H^1 = 0.4\%$ and $n_H^2 = =0.3\%$. The experiments were performed in a temperature range $80 \leqq T \leqq 373$ K for $0.1 \leqq q \leqq 1.9$ Å^{-1}. The measurements were made using the high-resolution backscattering spectrometer at the ILL (Institute Laue-Langevin) Grenoble. Thereby the quasielastic lines were investigated within an energy window of $\Delta E = \pm 5.3$ µeV.

As a consequence of the small scanning range ΔE, at larger q's only the narrow component could be measured. The integrated intensity of the quasielastic line, as observed within ΔE as a function of q and temperature is shown in Fig.4.11. The intensity has been normalized to the total intensity I obtained at 73 K. For small q nearly the full intensity is concentrated within the energy window. This demonstrates the existence of only one line which falls entirely into the instrumental energy range. However, at larger q's, only a fraction of the intensity appears in the integration window which decreases with increasing temperatures. This behavior reflects the reduction of the trapped fraction of protons with temperature.

Figure 4.12 presents the q dependence of the quasielastic width for the $n_N = 0.7\%$ sample at different temperatures, together with the corresponding curve for H in dilute solution in pure Nb; a dramatic decrease of linewidth due to impurity trapping is obvious. While in pure Nb the maximum of linewidth is reached around $q \cong 2$ Å^{-1}, in the doped sample such an extremum is reached already around $q \cong 1$ Å^{-1}. Finally,

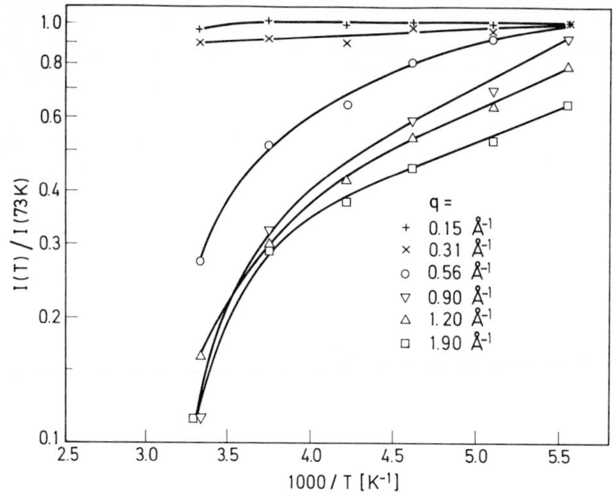

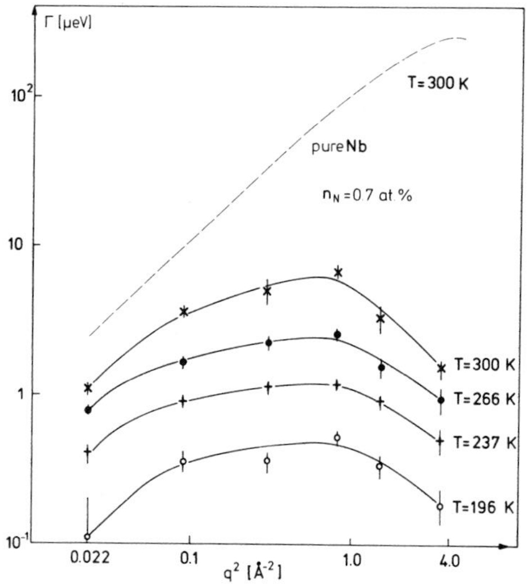

▲Fig.4.11. Intensities of the quasielastic spectra from the $NbN_{0.007}H_{0.004}$ sample within the energy window $\pm\Delta E$ as a function of inverse temperature for different q. The intensity is normalized to the total intensity determined at 73 K

Fig. 4.12. Linewidths of the spectra from $NbN_{0.007}H_{0.004}$ as a function of scattering vector q and temperature. (-----) H in pure Nb as extrapolated from /4.12/

deviations from the q^2 behavior (4.10) which in pure Nb is valid up to $q \cong 1$ $Å^{-1}$ (at least at 300 K, see Sect.2.2) occur already at the smallest q under consideration.

The scattering law of the two-state model contains all the features described above with the exception of the premature maximum in linewidth around $q \cong 1$ $Å^{-1}$, which presumably is caused by the internal structure of the trap and has been treated in detail by KEHR et al. /4.47/. Here we neglect all such structure and therefore limit the application of the two-state model to $q \leqq 0.9$ $Å^{-1}$. The experimental spectra for $q \leqq 0.9$ $Å^{-1}$ were fitted with the scattering law of the two-state model, τ_0 and τ_1 being the only disposable parameters.

For the undisturbed regions of the lattice, the diffusion coefficient of H in pure Nb was inserted. This assumption appears to be justified, since the long-range disturbances created by the impurities are small (typically in the order of 1 meV; see Sect.2.5.1) compared to the energy necessary to produce coincidence situations leading to diffusion (activation energy: 68 meV; see also Sect.2.2,3). Such small disturbances will probably not affect the incoherent diffusion mechanism.

The result of the fit is presented in Fig.4.13. Compared to typical jump rates τ^{-1} of 10^{+11} jumps/s, τ_0^{-1} and τ_1^{-1} are two orders of magnitude smaller. At higher temperatures $\tau_1 > \tau_0$ holds; the protons are mainly in undisturbed regions. At lower temperatures, we have $\tau_1 < \tau_0$, the protons are predominantly trapped. As a local property of the traps τ_0 does not depend on the N concentration, justifying the assumption of well-separated trapping and undisturbed regions in the lattice. From the activation energy of $1/\tau_0$ (166 meV) a binding energy of approximately 100 meV for the N-H pair can be deduced in agreement with resistivity /4.39/ and internal friction results /4.41/.

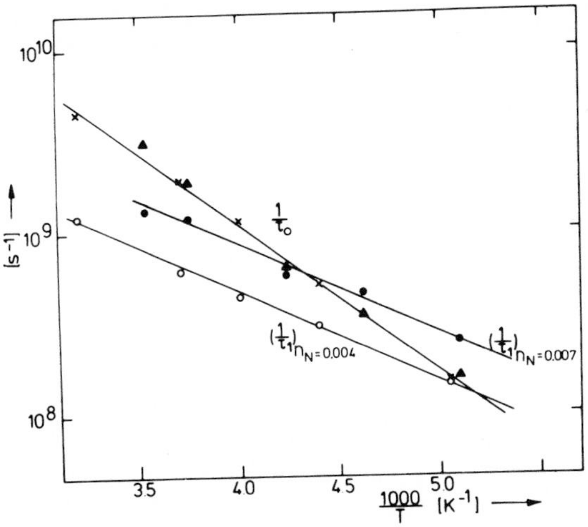

Fig.4.13. Escape rate $1/\tau_0$ and trapping rate $1/\tau_1$ as a function of inverse temperature and nitrogen content. For $1/\tau_0$ (▲) n_N = 0.7%, (x) n_N = 0.4%

The capture rate $1/\tau_1$ depends on the N concentration. The ratio of $1/\tau_1$ for the two N concentrations of 1.8 ± 0.2 compared to the ratio for N concentrations of 1.9 ± 0.2 is in agreement with theory. According to (2.31) the activation energy for $1/\tau_1$ should agree with the activation energy for the self-diffusion coefficient D. The resulting higher value (94 meV) probably originates from saturation effects of the energetically most favorable site close to the impurity: at low temperatures each N impurity can only prevent *one* proton from precipitating into the hydride phase /4.39/. After correction for this effect, (2.31) allows evaluation of the trapping radius r_t. Values of the order of 5 Å result. They are in good agreement with calculations using anisotropic continuum theory. Furthermore, the mean free path between two trapping events can be calculated. It decreases with increasing temperature, demonstrating partial saturation of the traps. Its absolute value is slightly larger than the mean distance between the nitrogen impurities.

By using the values of τ_0 and τ_1 and the diffusion coefficient in pure Nb according to (4.36) the effective self-diffusion coefficient can be evaluated. It closely agrees with the values obtained from the linewidth at small q (q = 0.15 Å) which are shown in Fig.4.14 for n_H = 0.7%. In contrast to pure Nb, where the activation energy changes around 250 K (Table 4.1), an Arrhenius relation holds over the entire T range.

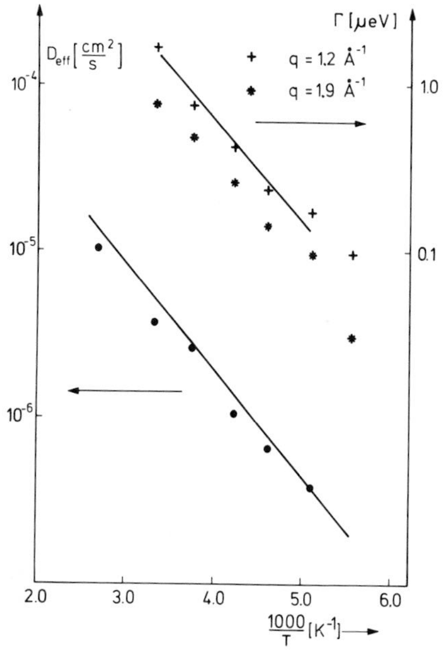

Fig.4.14. Effective diffusion coefficient (lower left part) and width of the quasielastic spectrum for two q values (upper right part) as a function of inverse temperature. The lines are the result of theoretical calculations /4.45/

According to the two-state model, at large q values a direct determination of the escape rate $1/\tau_0$ should be possible. Figure 4.14 presents the width of the narrow components in the quasielastic spectrum for two large q values. The activation energy for $\Gamma(T)$ at q = 1.9 $Å^{-1}$ is only 115 meV, considerably smaller than the expected value of ∼170 meV (binding energy ∼100 meV plus energy of migration ∼70 meV). Thus, the connection between $\Gamma(T)$ and the escape process is more intricate than expected.

The trapping radius r_t = 5 Å and calculations of the elastic interaction energy between N and H as a function of distance and orientation in the lattice /2.39/ show that many sites around the impurity are energetically disturbed, leading to a *trapping region* with a spectrum of site energies. The influence of trapping regions on the diffusional process shows some characteristic differences from that of single deep traps: (i) The long-range diffusion coefficient D_{eff} is affected by a distribution of activation energies rather than by an escape rate exhibiting a simple activated behavior. In the absence of drift motion we have $D_{eff} = D/Z$, where $Z = \sum_n \exp(-\beta E_n)$ (4.23) is the partition function of the protons with respect to the site energies E_n /2.41/. In this case, the temperature effect of the impurities on the diffusion coefficient is much smoother than expected for a single deep trap. (ii) At intermediate temperatures, where trapping does not yet compete with separation into hydride phases, saturation should be much less pronounced than in a model where each impurity atom provides only one deep trap. Therefore, changes in the H concentration at a given impurity content will have only a minor influence on the diffusion coefficient as revealed by recent Gorsky effect measurements /4.48/. (iii) A spectrum of site energies leads to a distribution of different H jump rates within the trapping region. This will affect the scattering function $S^D_{inc}(\underline{q},\omega)$ at large momentum transfers, where the single jump rates enter. $S^D_{inc}(\underline{q},\omega)$ then is the envelope of Lorentzians with widths related to the different jump rates (4.24), each of which exhibits a different temperature behavior. The resulting temperature dependence of this envelope may thereby be very different from that of the single jump rates /2.41/.

In order to study this problem more quantitatively, a model was investigated which explicitly takes into account the lattice distortions created by the nitrogen interstitials /4.45,49/. The nitrogen atoms were located at tetrahedral sites in a niobium lattice.[6] The lattice distortion was calculated using the Kanzaki forces of the nitrogen atoms, obtained from lattice expansion and Snoek effect measurements, and the niobium force constants, determined from phonon spectra. The equilibrium energies were determined from the resulting elastic H-N interaction. The potential maxima were assumed to be equal. Electronic contributions to the hydrogen-nitrogen interaction were approximated by a hard-core potential. The self-correlation function for diffusion of a proton, $G_s(\underline{r},t)$, was calculated by numerically solving the rate equations. A cube with 16,000 niobium atoms and 128 nitrogen atoms was used. The results for D_{eff} and the width of the narrow component for large q are shown in Fig.4.14 as two lines.

[6] Today it is known that N occupies O sites in the Nb lattice. The local strain fields very close to the impurity, therefore, may differ from those calculated in this model. However, since even at very short distances, lattice and continuum theory reveal nearly identical results /2.5/, the actual site occupation of the N has only a minor influence on the site energy distribution away from the nearest neighborhood of the impurity. Thus it is expected that the characteristic features of the model calculation will not depend strongly on the actual N site.

Two features are noteworthy: (i) Though the model has only one disposable parameter – the hard-core radius around the impurity – it describes the magnitude of the observed linewidth at large q correctly (it is reduced by more than a factor of 100 compared to pure Nb), and it also yields the observed activation energy. Thus, the observation of a low apparent activation energy at large q appears to be indicative of an extended trapping region containing a spectrum of different jump rates. (ii) At the same time, the calculated effective diffusion coefficient D_{eff} agrees well with the experimental data. In particular, it reproduces also the simple activated behavior of the diffusion coefficient, which contrasts with that in pure Nb, where at 250 K the activation energy changes from 106 meV to 68 meV. Thus, the apparent Arrhenius behavior of D_{eff} can be understood as a compensation of the decreasing slope of ln D versus 1/T by an increasing efficiency of the traps /2.39/. (iii) Within this model the above-mentioned intensity of D_{eff} to n_H /4.48/ can also be understood: the protons block only the energetically deepest sites, whereas the change in D is determined by the totality of the impurity-affected sites.

Obviously, the phenomenological two-state model leads to results very similar to those obtained from the microscopic model described above. In particular it explains the disappearence of the change in activation energy caused by impurities. VÖLKL and ALEFELD /4.50/ and more recently QI et al. /4.48/ suggested that this observation results from an impurity-induced suppression of a fast H jump mechanism at low temperatures. Following the argument (ii) given above, such an assumption is not required. In this context we should point out that the observed ratio D_{eff}/D for a given temperature and nitrogen concentration is nearly the same for both H and D /4.48/. This is expected from the relation $D_{eff} = D/Z$ /2.41/, where isotope effects can only appear in the diffusion coefficient.

Finally, with regard to the limits of application for the two-state model in a system with trapping regions, it is obvious that the approximations of the two-state model are only valid as long as the trapping regions are well separated and do not affect more than a few percent of the total lattice sites. With increasing impurity concentration and with decreasing temperature the trapping regions grow, and ultimately the two-state model has to be replaced by a concept of diffusion in a disordered system with random energy shifts affecting its lattice sites.

The influence of substitutional impurities on H diffusion in Nb has been investigated on the system $NbV_{0.002}H_{0.004}$ by quasielastic neutron scattering and preliminary results have been reported /4.45,51/. The quasielastic spectra were measured in the temperature region 178 K $\leq$ T $\leq$ 247 K for q's 0.13 $\leq$ q $\leq$ 1.8 $Å^{-1}$. Figure 4.15 shows the width Γ of the spectra for the two smallest q values as a function of inverse temperature. Assuming that in this q range the linewidth is still proportional to the effective H diffusion coefficient of the system, D_{eff} was evaluated. For compar-

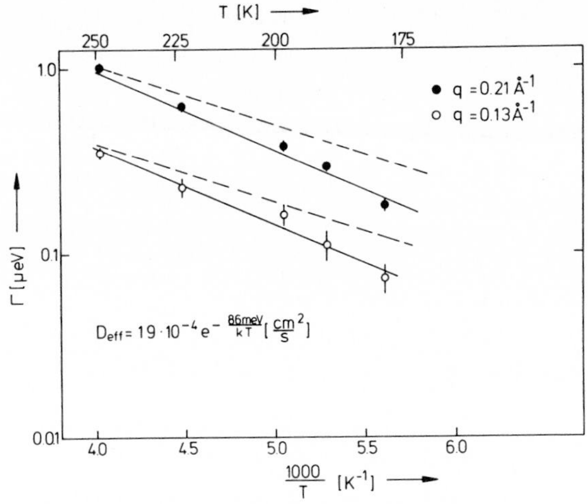

Fig.4.15. Quasielastic linewidth of H in $NbV_{0.002}H_{0.004}$ at small q. (-----) linewidth expected for a pure NbH sample /4.12/

ison the temperature dependence of the linewidth for a pure Nb sample /4.12/ is indicated as a dashed line. The activation energy for H diffusion in this V doped sample increased from 68 meV in pure Nb to 86 meV.

In the next step of data refinement, all spectra with $q \leq 1$ Å^{-1} (see above) were fitted simultaneously with the scattering law of the two-state model (4.31). Thereby an exponential temperature dependence of the two rates $1/\tau_0$ and $1/\tau_1$ was assumed. The activation energy for $1/\tau_1$ was taken equal to the energy of migration in the undisturbed lattice (68 meV) as explained in the case of NbN. Under these circumstances the fitting process yields an activation energy for $1/\tau_0$ of $E_a = 163 \pm 10$ meV. From E_a a binding energy $E_b = 95 \pm 10$ meV can be derived. This value is in agreement with an NMR result of 90 ± 50 meV /4.52/. The evaluated prefactor for $1/\tau_1$ of $\tau_{10}^{-1} = (3.6 \pm 0.3) \times 10^{10}$ s^{-1} corresponds to a trapping radius of $R_t = 3.2$ Å. Due to the very low impurity concentration, however, data evaluation at large q was impossible and information on H dynamics in the trapped state could not be obtained.

4.2.4 Hydrogen Diffusion in Simple Concentrated Systems

With increasing hydrogen concentration a finite probability exists that interstitial sites neighboring a given H atom are blocked by the presence of other H atoms leading to correlations between consecutive jumps. For instance, if a particle has performed a jump, then with certainty a vacancy is behind it and hence the probability for a back jump is enhanced. For arbitrary concentrations these H correlations constitute a serious problem for the theoretical calculations of the respective correlation functions which have been treated mainly by Monte Carlo techniques /4.53-56/.

Before we discuss the Monte Carlo results we shall outline concepts for the case of nearly saturated systems ($n_H \to 1$).

For limiting high concentration H diffusion is closely related to the dilute case, since it can be treated exactly as a random walk of dilute vacancies. The importance of this mechanism for tracer diffusion in metals has stimulated thorough investigations which have been reviewed, e.g., by LE CLAIRE /4.57/. Two concepts are commonly put forward.

(i) The encounter model /4.58/ considers occasional encounters of the tracer atom with a vacancy.

The interaction time τ_i of the vacancy with one particular atom is assumed to be very short compared to the time between encounters with different vacancies, τ, which is also the mean rest time at a certain lattice site. During the interaction time several rapid site exchanges between the vacancy and the considered atom may take place. In the scattering function these processes appear only for energy transfers $\hbar\omega \sim \hbar/\tau_i$ far away from the central part which is characterized by $1/\tau$. Neglecting these wing contributions, the diffusional process can be modeled again in terms of a Chudley-Elliott model, where the different not correlated encounters take over the role of the independent jumps in the dilute case. If we now replace the spatial H distribution after one jump $f(\underline{q})$ (4.3,4) by

$$\tilde{f}(\underline{q}) = 1 - \sum_{i=1}^{n} w_i e^{i\underline{q}\underline{s}_i} \quad , \tag{4.35}$$ [7]

where w_i is the probability of motion from $\underline{r} = 0$ to the lattice point $\underline{s}_i$ during one encounter, then the incoherent scattering law can be calculated as in the preceding section for dilute concentrations. Assuming that besides blocking nothing else changes at higher concentrations, the time between different encounters $\tau(n_H)$ is given in mean-field approximation by

$$\tau(n_H) = \frac{\tau}{1-n_H} \quad . \tag{4.36}$$

For the incoherent scattering law we get in analogy to (4.6)

$$S^D_{inc}(\underline{q},\omega) = \frac{1}{\pi} \frac{\tilde{f}(\underline{q})/\tau(n_H)}{[\tilde{f}(\underline{q})/\tau(n_H)]^2 + \omega^2} \quad . \tag{4.37}$$

[7] Note that even for a jump model comprising next-neighbor jumps correlation effects can bring about effective transfers over larger distances. Thus, in general the sum is not restricted to next-nearest neighbors.

As outlined above, (4.37) is an approximation valid for ω smaller than the inverse interaction time $1/\tau_i$. For larger ω the details of the interaction process including the backward jump will appear in the scattering function. Encounter models for bcc and fcc metals have been treated in detail by WOLF /4.58,59/.

(ii) The correlation factor model /4.60/ takes into account the correlations between successive jumps introducing a reduction factor $r(n_H)$ for the tracer diffusion coefficient

$$D = \frac{r(n_H)d^2}{6\tau(n_H)} \quad , \tag{4.38}$$

$r(n_H)$ is defined by the expression

$$r(n_H) = \frac{1 + \langle\cos\theta\rangle}{1 - \langle\cos\theta\rangle} \quad , \tag{4.39}$$

where $\langle\cos\theta\rangle$ is the average over the angles between consecutive jump vectors and is subject to concentration-dependent correlations. The correlation factor and encounter models are related by

$$r(n_H{=}1) = \sum_{i=1}^{n} w_i \frac{s_i^2}{zd^2} \quad , \tag{4.40}$$

where z is the average number of jumps per encounter. From (4.35) in connection with (4.40) it is clear that the correlation factor is also q dependent.

Calculations for fcc metals have shown that it decreases slightly with increasing q /4.56/.

For arbitrary concentrations the encounter model is in difficulties, since the condition of uncorrelated encounters is not fulfilled any more, and Monte Carlo calculations appear to be the most convenient method of dealing with the problem in the general case. Up to now only fcc structures have been considered. ROSS and WILSON report that $S_{inc}(q,\omega)$ changes smoothly from the Chudley-Elliott form at $n_H \sim 0$ to the vacancy form for $n_H \sim 1$ /4.55/. At intermediate H concentrations small deviations from the Lorentzian shape occur. In particular, it is emphasized that two regions in ω have to be distinguished: for large ω, $[\omega \sim 1/\tau(0)]$ $S^D_{inc}(q,\omega)$ follows the mean-field description (the appropriate jump rate is $1/\tau(n_H)$ while for small ω $[\omega < 1/\tau(n_H)]$ the width is further reduced to $r(q,n_H)f(q)/\tau(n_H)$ yielding the correct diffusion coefficient at small q (4.38). In order to give an impression how the incoherent scattering law behaves as compared to the limiting forms at small and large ω, Fig.4.16 presents the Monte Carlo result of ROSS and WILSON /4.55/ for $q = (2\pi/a,0,0)$ and $n_H = 0.7$. As it is evident from Fig.4.16, the three curves exhibit only marginal differences. In particular a Chudley-Elliott scattering law

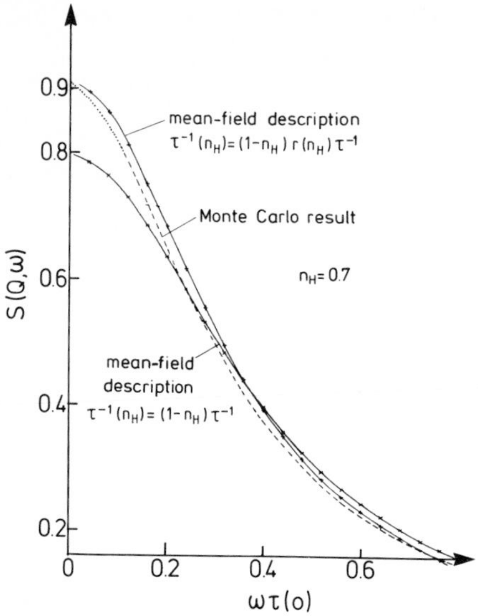

Fig.4.16. Plot of $S_{inc}(q,\omega)$ at the zone boundary $(2\pi/a,0,0)$ in a fcc lattice for $n_H = 0.7$. The large and small ω [$\tau(n_H)^{-1} = (1-n_H)\tau^{-1}$ and $\tau(n_H)^{-1} = (1-n_H)\tau^{-1}r(n_H)$] approximations are compared with the Monte Carlo result

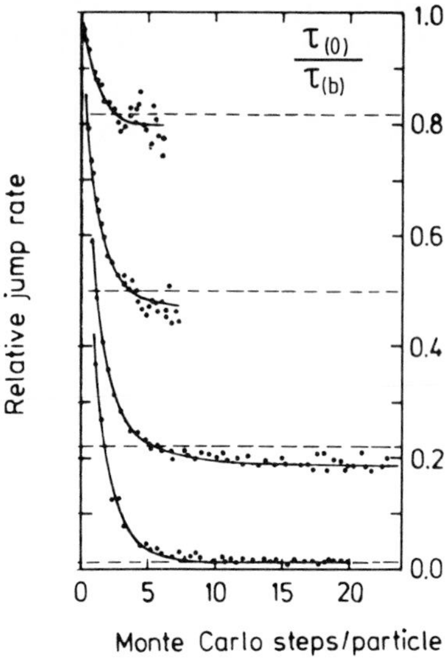

Fig.4.17. Normalized conditional backjump rates $1/\tau_b/1/\tau(0)$ for $n_H = 0.182$, 0.498, 0.777, 0.988 (from above); (-----) mean-field values

with a width function corrected for the q dependent correlation factor and blocking constitutes a very good approximation in the experimentally accessible range.

There remains the question of how the local correlations including the backjump show themselves in the scattering law. KEHR et al. /4.56/ have addressed this problem on the level of correlation functions, where the time-dependent correlations between consecutive jumps of a given H atom are visible. In particular, the backward correlation which is strong immediately after one jump has occurred disappears for longer times as the vacancy is filled by other particles. This time-dependent behavior of the backward correlation can be demonstrated most clearly in terms of so-called waiting-time distributions $\psi(t)$ for the particle after it has performed a jump. Thereby $\psi(t)dt$ is the joint probability that the particle has made no jump until time t and performs a jump between t and $t+dt$. The waiting-time distributions have been calculated for fcc lattices and various H concentrations by Monte Carlo techniques. Figure 4.17 presents results for the conditional backjump rate $1/\tau_b$ normalized to $1/\tau(0)$, which is related to the waiting-time distribution and is the rate for a backjump under the condition that no jump has occurred until time t. As it can be seen, immediately after the initial jump, the backjump rate $1/\tau_b$ is equal to the jump rate in the empty lattice $1/\tau(0)$. With increasing time the enhanced backjump rate decays rapidly and falls slightly below the mean-field rate $1/\tau(n_H)$, in order to compensate for the faster initial rate.

Qualitatively we expect that the fast backjump rate appears in the incoherent scattering law at large momentum transfers and sufficiently large ω. Quantitatively, however, its influence on $S^D_{inc}(q,\omega)$ has not yet been analyzed. Considering the large coordination number z = 12 for the fcc structure, it is not a favorable case for a study of backward correlations. The number of channels through which the vacancy behind an atom is filled after a jump is large and also the possibilities for forward jumps are numerous. Lattices with lower coordination numbers like the interstitial lattice in the bcc structure (z = 4) promise to show much larger effects.

In view of the above discussion it is no surprise that neutron scattering experiments on H diffusion in β-$PdH_{0.73}$ /4.61/ agreed well with a simple Chudley-Elliott model /4.6/. Also deviations from a Lorentzian line shape were not obvious. The obtained quasielastic widths are shown in Fig.4.18 for three different temperatures and are compared with the prediction of the CE model. Thereby the mean-field value $1/\tau(n_H)$ for the jump rate was used which was extrapolated from low-concentration measurements. The agreement is excellent and we wonder whether the 10% - 15% downward correction, expected from the correlation factor, has been lost in the multiple scattering correction procedures or whether the increasing H concentration also influences the attempt jump rate $1/\tau(0)$ itself. More recent evaluations of these spectra including more measurements up to 200 °C allowed for the change of the scattering function between both asymptotic ranges introducing an "elastic" fraction, in order to account for the narrow central part of the spectra. The procedure was applied to the experimental spectra as well as to the Monte Carlo results and appears

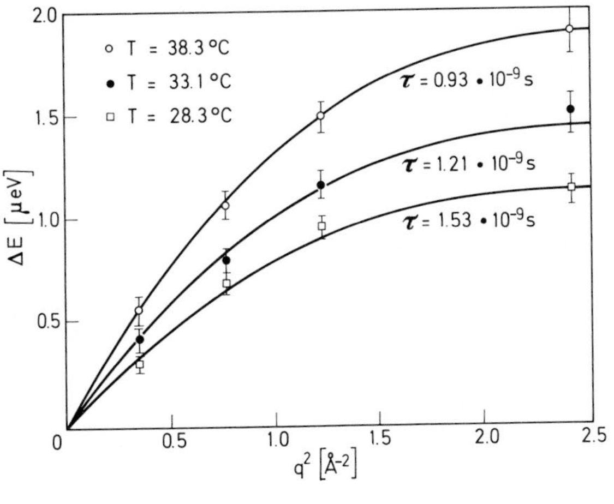

Fig.4.18. Quasielastic linewidths ΔE obtained from $PdH_{0.73}$ as a function of q^2. (——) predictions of a CE model setting τ to the mean-field value $\tau = \tau(0)/(1-n_H)$

to have removed the above difficulty. The authors claim that Monte Carlo calculation and experiment agree well and yield the $\tau(0)$ values observed for infinite dilution /4.62/.

Clearly, more experimental work is needed in this field, which we advise should concentrate on bcc hydrides. There, from the lower coordination number ($z = 4$) we expect more pronounced effects in the scattering function.

4.2.5 Hydrogen Diffusion in Potential H Storage Materials

Other than the "classic" monoatomic metal hydrides like NbH or PdH, metal hydrides for hydrogen storage purpose are in general binary alloys like FeTi, $LaNi_5$, Mg_2Ni, $Ti_{1.2}Mn_{1.8}$, or even more complicated structures like $Ti_{0.8}Zr_{0.2}CrMn$. Since the intermetallics are composed of different metal atoms, chemically different H sites exist as a consequence of the varying H affinities of the host atoms, e.g., in $Ti_{1.2}Mn_{1.8}$ tetrahedrons are formed from 4 Mn atoms, 3 Mn and Ti, and 2 Mn and 2 Ti, the latter containing the most attractive H site because of the large H affinity of Ti. In addition, the crystal structures in general are complicated, exhibiting large unit cells with numerous crystallographically inequivalent sites. Thus, contrary to the monoatomic hydrides, a whole spectrum of energetically different H sites can be expected, and H diffusion in such structures comprises aspects of H trapping and of correlated diffusion due to the high hydrogen concentrations. Finally, it is mostly impossible to produce single crystalline samples because the alloys tend to break into powder upon hydrogenation. In this situation it is virtually impossible to investigate hydrogen motion in such detail as, for instance, in Nb or Pd.

Therefore, most of the neutron-scattering experiments on H storage materials were confined to small momentum transfers, where the H diffusion coefficient can be determined irrespective of the geometrical details (4.10,34). Here, QNS is in particular useful because it measures the macroscopic diffusion coefficient as a true bulk experiment unaffected by inner and outer surfaces. At the same time most of the macroscopic techniques like Gorsky effect are not applicable due to the powder morphology. Other microscopic methods like NMR except for the so-called pulsed field gradient techniques /4.63/ measure only single relaxation times, the interpretation of which in terms of diffusion coefficients is very difficult if a whole spectrum of times is present. Compared to the bcc hydrides or also to Pd, the diffusion coefficients were found to be small, and QNS can be applied only using high-resolution techniques like neutron backscattering /4.64/.

High-resolution, low q QNS diffusion experiments have been reported for FeTiH /4.65/, Ti_2NiH_2, $LaNi_5H_6$ /4.66-68/, and $Ti_{1.2}Mn_{1.8}H_3$ /4.69/. Figure 4.19 presents

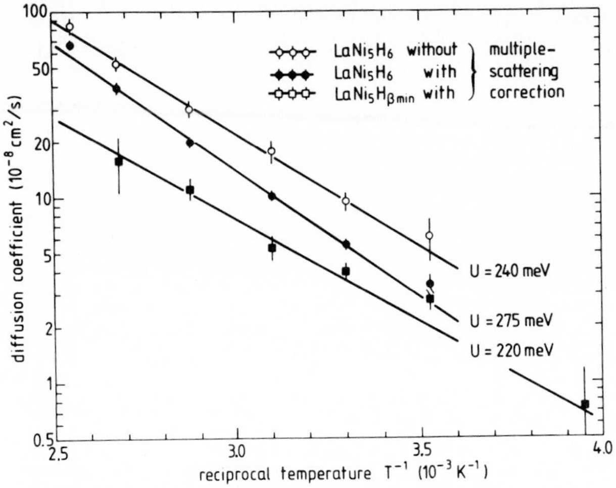

Fig.4.19. H diffusion coefficient in $LaNi_5H_6$ before and after multiple-scattering correction. Included are also results taken at a lower H concentration yielding a *smaller* diffusion coefficient

recent QNS results on the H diffusion coefficient in $LaNi_5H_6$ /4.68/ and emphasizes the importance of multiple-scattering corrections for such experiments. Multiple-scattering processes tend to increase the linewidth at small momentum transfers, since two-times scattered neutrons are mainly scattered twice at large q, where the linewidths are broader. Their influence on the apparent width increases the more the multiple scattered neutrons contribute in an energy range defined by the resolution of the spectrometer, and is largest at low temperatures. Thus without corrections the observed activation energy tends to be smaller than the real one. For $LaNi_5H_6$ with a neutron transmission of 80%, which is typical for such experiments, the activation energy changes by ~15% after correction.

Recently, HEMPELMANN et al. obtained more microscopic information on the H diffusion process for the example of $Ti_{1.2}Mn_{1.8}$ /4.69/. Thereby the main features of the motional mechanisms in this complex structure were described adopting the two-state model for diffusion in the presence of traps as outlined in Sects.2,4 and 4.2.3. This description was motivated from the observation of composed spectra whose components exhibit q and T dependencies characteristic for diffusion in the presence of traps: Figure 4.20 presents the q dependence of the linewidths for the two components as observed at T = 355 K together with their T dependence in form of an Arrhenius plot. For q → 0 only the linewidth of the narrow component tends to zero whereas at large q the broad component has a pronounced q dependence. The Arrhenius representation of the linewidths reveals a considerably higher activation energy for the narrow component. Figure 4.21 displays the q dependence of the

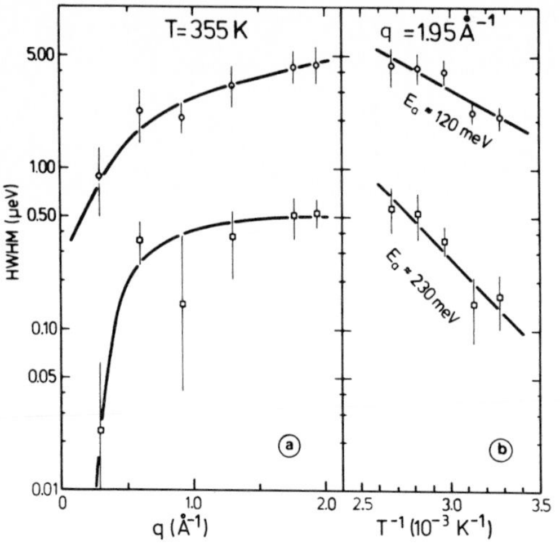

Fig.4.20. q and T dependence of the quasielastic linewidths obtained from a fit of two Lorentzians to the experimental spectra taken on $Ti_{1.2}Mn_{1.8}H_3$

▼Fig.4.21. q and T dependence of the intensities of the two Lorentzians of Fig.4.20 relative to the total scattering intensity obtained at 101 K

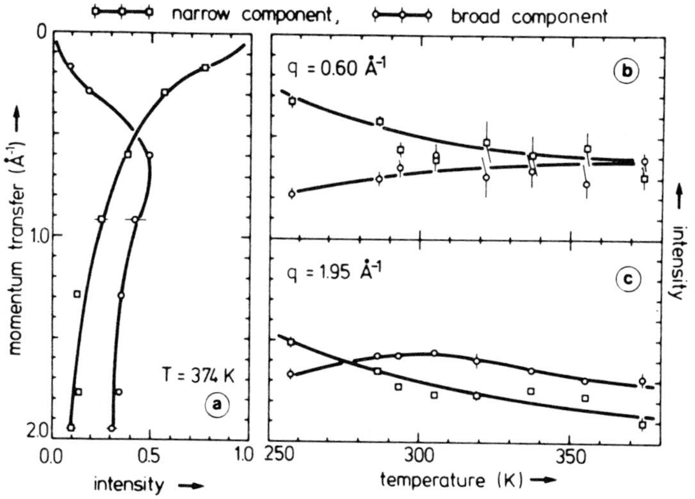

relative intensities of both components relative to the total scattering intensity at 101 K. Starting from a weight close to 1 at q = 0 the weight of the narrow line continuously decreases with increasing q. The relative intensity of the broad component on the other hand starts from weight zero at q = 0, passes through a maximum around q ≅ 0.8 and decreases again towards higher q. At large q, the total intensity of both components is significantly lower than expected from the reference. The observed q-dependent weights closely resemble the theoretical curves of the two-state model (Fig.4.10).

As discussed above, in intermetallic compounds, a whole spectrum of energetically different H sites is expected. In order to account for this feature, we divide the available sites into 'trap sites', comprising the energetically lowest interstitial positions, and 'free sites' otherwise. Dissolved H occupies preferentially the trap sites and according to the thermal occupation probability saturates most of them. The remaining H is distributed over the free energetically less favorable sites and occasionally gets trapped in an empty trap site. Thus, in spite of the high density of traps, for a single H atom, the diffusional process with respect to trapping is not very different from the situation in Nb with dilute N impurities, since the empty traps are also dilute. In order to distinguish from impurity traps, we call the energetically favorable sites in intermetallics structural traps.

According to Fig.4.21 a substantial intensity fraction is missing at large q, indicating the existence of broad components in the spectrum the intensity of which lies mostly outside the energy range of the spectrometer ($\Delta E = \pm 10$ μeV). Such a third component in the spectrum with a substantial weight only at large q must be related to a rapid motion confined to small regions in space. Two kinds of such a local motion are imaginable: (i) correlated jumps as a consequence of the high hydrogen concentration (Sect.4.2.4) and (ii) structural effects like local hopping in an extended trap as has been observed for $NbO(N)_xH_y$ (Sect.4.4).

In order to account for this local motion, the two-state model was extended to a three-state model. In order to model correlated jumps which occur during passages of "free diffusion", the local motion was incorporated into the self-correlation function of the free state $[G_{f0}(\underline{r},t)]$ (Sect.4.2.3). Alternatively, in order to describe local hopping at a trap, this process was connected with the self-correlation function in the trapped state $[G_{t0}(r,t)]$ (4.25). The spatial extension of the local jump process was considered as a dumbbell of length ℓ; the corresponding rate is $1/\tau$. While the significance of the parameters $1/\tau_0$ and $1/\tau$ remains unchanged, some care has to be taken with respect to the trapping rate $1/\tau_1$, where saturation effects explicitly have to be included. Equation (2.31) changes into

$$\tau_1^{-1} = 4\pi r_t D\, n_t^*(T) \quad , \tag{4.41}$$

where $n_t^*(T)$ is the concentration of empty traps. The latter can be inferred from thermodynamic arguments considering the chemical potentials of free and trapped hydrogen atoms, respectively, which in thermal equilibrium have to equal /4.69/

$$kT\ln\left(\frac{z_f/Z_f}{1-z_f/Z_f}\right) = -E_b + kT\ln\left(\frac{z_t/Z_t}{1-z_t/Z_t}\right) \quad , \tag{4.42}$$

where E_b is the binding energy at the structural trap, z_f and z_t are the number of protons in the "free" and "trapped" states, respectively, and Z_f and Z_t are the numbers of these sites available. After some algebra the concentration of empty traps is obtained

$$n_t^* = \frac{1}{2}\left(n_t + n_H + \frac{\exp(-E_b/kT)}{1-\exp(-E_b/kT)}\right) \pm \frac{1}{2}\left[\left(n_H + n_t + \frac{\exp(-E_b/kT)}{1-\exp(-E_b/kT)}\right)^2 - \frac{4n_H n_t}{1-\exp(-E_b/kT)}\right]^{1/2} \quad (4.43)$$

where n_t is the total trap concentration and n_H is the hydrogen concentration. The positive sign is valid for $n_H > n_t$ while the negative sign applies to $n_H < n_t$. The experimental spectra obtained for 8 temperatures between $228 \leq T \leq 374$ K and for $0.17 \leq q \leq 1.95$ Å^{-1} were fitted simultaneously with this three-state model associating the local relaxation either with the free or with the trapped state. Thereby, significantly better agreement was achieved relating local hopping to the free state. The q and T dependence of the resulting linewidths is shown in Fig.4.22. With respect to the q dependence a close relation to the theoretical curves for the two-state model (Fig.4.10) is obvious. The activation energies for τ^{-1}, τ_ℓ^{-1}, and τ_0^{-1} are 210 meV, 209 meV, and 300 meV, respectively. While the trap concentration n_t is evaluated to $n_t = 0.24 \pm 0.06$, the concentration of free traps n_t^* varies between 5% and 9% depending on temperature which is consistent with the application of (4.41). The relatively large amount of trapping sites (25%) clearly shows that they do not originate from impurities or surface segregations. An experiment on stoichiometric $ZrMn_2H_3$ revealed analogously composed QNS spectra and thus the structural traps do not appear to be a peculiarity due to the excess Ti atoms. Finally, the jump length of the local jump process was determinated to $\ell = 1.37$ Å in fair agreement with the nearest-neighbor distance of 1.3 Å. Putting together the evidence: the local-jump process is associated with the diffusional state, its activation energy is closely

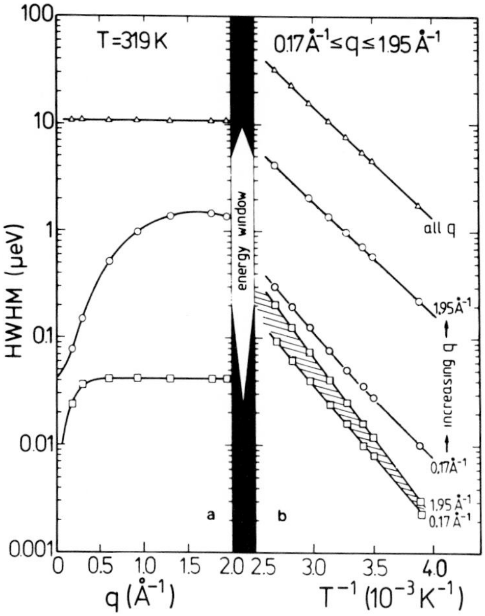

Fig.4.22. q and T dependence of the widths of the three lines in the QNS spectra of $Ti_{1.2}Mn_{1.8}H_3$. These widths have been calculated with the parameters resulting from a global fit with the three-state model to 48 experimental spectra including multiple-scattering corrections. The points indicate at which q and T values the spectra were recorded

related to that of the free state, its jump length is given by a next-neighbor distance — it was suggested that the rapid local motion is connected with correlation effects due to high H concentrations as discussed in the previous section. Similar phenomena have also been observed in $LaNi_5H_6$ /4.68/ and $Ti_{0.8}Zr_{0.2}CrMn$ /4.70/.

The QNS study on $Ti_{1.2}Mn_{1.8}H_3$ is the first of its kind and appears to have uncovered significant microscopic details of H diffusion in concentrated H storage materials. Of particular importance is the existence of H sites acting as structural traps. They can cause various anomalous features as, for instance, a reverse concentration dependence of the H diffusion coefficient as recently observed in $LaNi_5H_x$ /4.68/ (Fig.4.19), where similar composed spectra have been seen at large q.

4.3 Phonons and Hydrogen Diffusion

During its rest time at a particular interstitial site, the proton exhibits vibrational motion around its equilibrium position. Two types of motion have to be distinguished.

(i) Localized vibrations of the H atom against its metal neighbors with frequencies typically a few times higher than the acoustic frequencies of the host lattice occur. They yield information on the strength of the metal-hydrogen interaction and therewith on the hydrogen potential. Its knowledge is of great importance for any quantitative evaluation of hydrogen diffusion properties. Recent precision measurements of the local vibrations of H, D and T including their higher harmonics in Nb and Ta will be surveyed.

(ii) In the so-called band modes the hydrogen moves in phase with the acoustic vibrations of the host. The resonant-like enhancements of the hydrogen amplitude in $NbH_{0.05}$ for certain lattice modes and their possible relation to lattice-activated H diffusion in Nb will be discussed.

4.3.1 Local Vibrations of Hydrogen

The investigation of hydrogen vibrations in metals is nearly as old as the method of inelastic neutron scattering, and a large amount of experimental data has been compiled. For an overview we refer to a recent review /4.71/. Here we concentrate on new developments concerning the local vibrations of H and its isotopes in bcc Nb and Ta which allowed a detailed determination of the hydrogen potential including anharmonic corrections.

Inelastic neutron cross section and the weakly anharmonic H potential

While in the fcc hydrides direct H-H interaction leads to considerable dispersion of the optical modes /4.72/ and inhibits a precise determination of the H potential in the hydride phase, the bcc hydrides like NbD_x exhibit virtually no q dependence of the vibrational levels. Even for high H concentrations collective H vibrations are of no importance, and the H isotopes can be regarded as independent single Einstein oscillators vibrating in their individual potential wells /4.73/.

In Nb and Ta the hydrogen isotopes occupy tetrahedral sites which in bcc metals have tetragonal symmetry. Consequently, the vibrational frequencies are split into two fundamentals, the upper one being degenerate ($\omega_1, \omega_2 = \omega_3$). Assuming longitudinal forces f to the next-neighbor metal atoms only, the dynamical matrix has the form

$$\underline{\underline{D}} = \frac{4}{5} f \begin{pmatrix} 2 & & \\ & 2 & \\ & & 1 \end{pmatrix} \tag{4.44}$$

and we have in addition $\omega_2/\omega_1 = \sqrt{2}$. Omitting acoustic-optic multiphonon processes, the double differential neutron cross section $\partial^2\sigma/\partial\omega\partial\Omega$ for scattering on a three-dimensional harmonic oscillator including all higher harmonics for energy loss processes is given by

$$\frac{\partial^2\sigma}{\partial\omega\partial\Omega} = \frac{\sigma^{tot} k_f}{4\pi k_i} \exp\left[-2W_b(q) - 2W_\ell(q)\right] \sum_{\substack{n,m,\ell=0 \\ n+m+\ell\neq 0}}^{\infty} \frac{1}{n!}\frac{1}{m!}\frac{1}{\ell!}$$
$$\times \left(\frac{\hbar q^2}{6M\omega_1}\right)^n \left(\frac{\hbar q^2}{6M\omega_2}\right)^m \left(\frac{\hbar q^2}{6M\omega_3}\right)^\ell \delta\left[\omega - (n\omega_1 + m\omega_2 + \ell\omega_3)\right] \quad , \tag{4.45}$$

where n,m,ℓ are the three vibrational quantum numbers and W_b and W_ℓ are the Debye-Waller factor contributions from the band and local modes, respectively. Since for all experimental temperatures $kT \ll \hbar\omega_i$ holds, only occupation of the ground state has been assumed. In (4.45) the total cross section σ^{tot} appears — for uncorrelated motions coherent and incoherent scattering both yield the density of states.

We now consider anharmonic disturbances of the hydrogen potential. The possible cubic and quartic anharmonic terms for H in a tetragonal surrounding (point group D_{2h}, $x = y \neq z$) can be found by group theory and have been calculated together with the corresponding corrections for the vibrational levels by ECKERT et al. /4.75/. Here we take a simpler approach and evaluate the anharmonic corrections for the one-dimensional potential $V(x) = a_2x^2 + a_4x^4$ in first-order perturbation theory, which already exhibits all important features of the detailed model. Its influence on the frequencies is twofold. (i) Relative to the ground state position the nth vibrational level is shifted by /4.76/

$$\Delta E^{H(D,T)} = \frac{3}{2}\frac{\hbar^2}{M^{H(D,T)}}\frac{a_4}{a_2}(n^2+n) \quad . \tag{4.46}$$

(ii) Since ΔE scales with 1/M, while the harmonic frequencies are proportional to $1/\sqrt{M}$, a_4 also causes deviations from the harmonic isotope effect

$$\frac{\hbar\omega^H}{\hbar\omega^{D(T)}} = \frac{\sqrt{2(3)}}{1-\Delta E^H/\hbar\omega^H\left[1-\frac{1}{2(3)}\sqrt{2(3)}\right]} \quad , \tag{4.47}$$

In general, a measurement of the isotope effect alone does not allow a distinction between an anharmonic potential which is identical for all isotopes and the case of different but harmonic potentials for H, D and T. According to (4.46,47), however, an additional determination of the frequency shifts ΔE provides the necessary information.

Different potentials for the H isotopes may result from a different electron density around the interstitial as a consequence of the significantly different zero point vibrations. In particular, an experimental decision between both possibilities would be extremely useful in the case of PdH_x, where anharmonic potentials as well as different electronic structures have been invoked to explain the inverse isotope effect in the transition temperature to superconductivity /4.77,78/.

Experimental results on Nb and Ta

The observation of a well-defined second harmonic to the lower deuterium frequency in Nb /4.74/ stimulated new interest in local mode investigations on H in bcc refractory metals. This section concentrates on new results on Nb and Ta which were performed applying the Be filter technique, where neutrons of varying high incident energies create phonons in the sample and are downscattered. Only those neutrons with energies below 5.2 meV can pass the Be filter, which is placed in front of the detector, and are counted. Since the filter cutoff is sharp and the transmission below the cutoff energy is high, at energies above 100 meV, the Be filter technique surpasses other neutron methods for the measurements of densities of state.

Figure 4.23 presents the spectrum obtained from $NbD_{0.85}$ at 10 K by RICHTER and SHAPIRO /4.74/ which for the first time showed the existence of higher harmonics to the D vibrations in a bcc metal. Besides the two peaks corresponding to the two fundamentals, a third peak evolves at 170 meV, 7 meV below twice the frequency of the first fundamental $\hbar\omega_1$.

Figure 4.24 presents spectra obtained by RUSH et al. /4.79/ from Nb containing H, D or T in the β phase at room temperature. While the $NbD_{0.72}$ and $NbH_{0.32}$ spectra con-

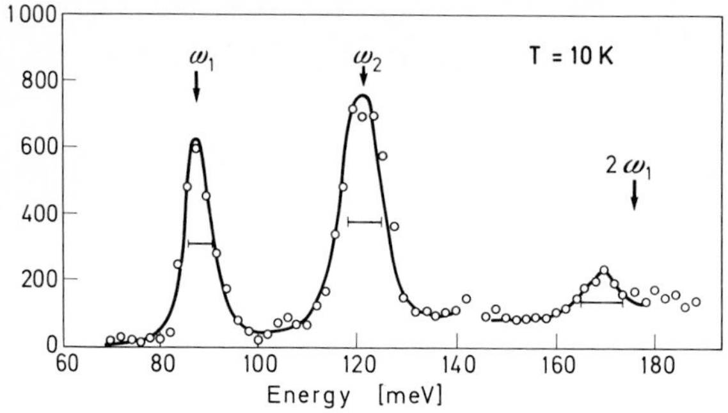

Fig.4.23. Inelastic spectrum obtained on $NbD_{0.85}$ at 10 K. The bars give the resolution width for each peak, the arrow $2\omega_1$ shows the position of the second harmonic if the potential would be harmonic

Fig.4.24. Inelastic spectra for $NbT_{0.2}$, $NbD_{0.72}$, and $NbH_{0.32}$ at 295 K

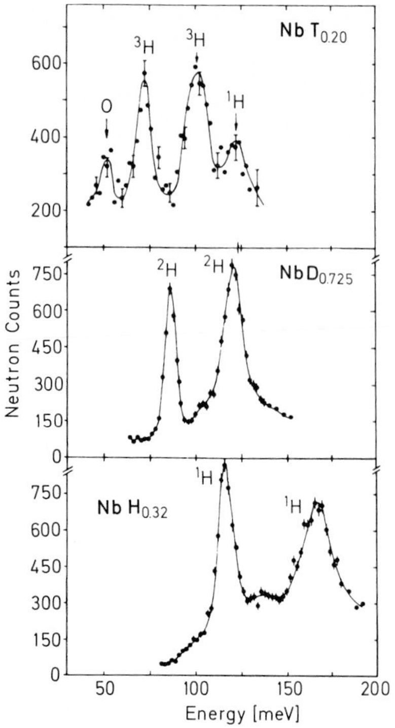

tain only the two fundamentals, the $NbT_{0.2}$ spectrum shows additional peaks which were associated with interstitial oxygen (52 meV) and a small proton fraction. It is interesting to note that the frequency of H within a tritium surrounding appears to be shifted to a considerably higher value (123 meV) compared to pure β-NbH_x (116 meV). The results for the peak positions are summarized in Table 4.3. As can be seen, the isotope shifts for the D and T vibrations with respect to the H vibration are significantly smaller than expected for an harmonic potential. Assuming an identical but anharmonic potential for all three isotopes, the frequency shifts expected for the second harmonics $\Delta E^H = 2\Delta E^D = 3\Delta E^T$ are evaluated from (4.47) and are compared with the measured value $\Delta E^H = 2\Delta E^D = -14$ meV for the lower fundamental in $NbD_{0.85}$ at 10 K. With respect to the experimental accuracy and the different temperatures of the two measurements, the agreement is satisfactory and indicates an interstitial potential common to all three isotopes. Concerning the second fundamental, measurements of second harmonics are not yet published but recent efforts at the neutron spallation source in Los Alamos revealed excitations in NbH_x up to 350 meV /4.75/. Their analysis will help to complete the picture of the H potential in Nb.

Figure 4.25 presents typical spectra obtained on $TaH(D)_x$ by HEMPELMANN et al. /4.80/. In the ordered β phase besides the two fundamentals again a third excitation appears at 228 meV in the hydride and at 168 meV in the deuteride. Both are

Table 4.3. Frequencies of the fundamental vibrations of hydrogen isotopes in Nb and Ta /4.74,79,80/ precipitated in hydride phases. ΔE_H is the energy shift of the second harmonic calculated from (4.47); ΔE_H^m is the measured frequency shift $\Delta E_H^m = 2\Delta E_D^m$ /4.74,80/

Isotope	Frequency	$\hbar\omega^H/\hbar\omega^i$	ΔE_H [meV]	ΔE_H^m [meV]
Nb	$\hbar\omega_1$ [meV]			
^1H	116 ± 0.7	1	-	-
^2H	86 ± 1	1.35 ± 0.02	-19 ± 6	-14 ± 1
^3H	72 ± 1	1.61 ± 0.0025	-21 ± 5	-
Nb	$\hbar\omega_2$ [meV]			
^1H	167 ± 1.5	1	-	-
^2H	120 ± 1.5	1.39 ± 0.01	-10 ± 4	-
^3H	101 ± 1	1.65 ± 0.02	-19 ± 5	-
Ta	$\hbar\omega_1$ [meV]			
^1H	121.3 ± 0.2	1		-16 ± 1
^2H	88.4 ± 0.4	1.37 ± 0.01	-13 ± 2	-18 ± 3
^3H	-	-	-	-
Ta	$\hbar\omega_2$ [meV]			
^1H	163.4 ± 0.4	1	-	-
^2H	118.7 ± 0.6	1.38 ± 0.01	-14 ± 3	-
^3H	-	-	-	-

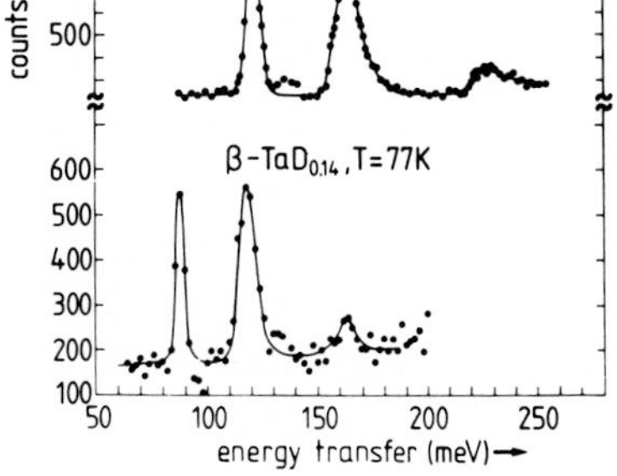

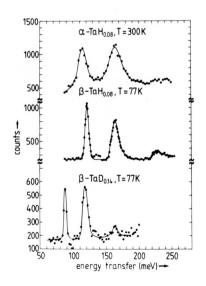

Fig.4.25. Vibrational spectra from H and D in Ta

shifted only slightly from the harmonic value of $2\hbar\omega_1$. The peak positions and the energy shifts are listed in Table 4.3. Applying (4.47) in the same way as for NbH(D,T) the energy shifts ΔE^H corresponding to the observed isotope effect in Ta have been evaluated and are compared with the measured shifts of the second harmonics of the lower fundamentals in the hydride and the deuteride. The agreement of both numbers is remarkably good and shows again that H and D oscillate in a common weakly anharmonic potential.

Finally, we comment on the implications of the observed deep interstitial potential wells on the possible diffusion mechanisms for H isotopes in bcc refractory metals. The existence of well-defined higher harmonics at energies well above the activation energy for H diffusion in these materials (106 meV for H in Nb, 140 meV for H in Ta /4.5/) clearly excludes simple classical diffusion across a barrier provided by the vibrational potential. Diffusion may occur by lattice-activated processes, where the phonons of the host lattice 'open the door' for H motions between interstitial sites. Classically, the vibrational configuration of the host atoms which leads to a jump of the interstitial is independent of the respective H isotope. The only isotope effect in the activation energy for diffusion arises from the different zero point vibrations of H, D and T and could qualitatively explain the observed experimental high-temperature diffusion data /4.81/. Quantum-mechanical jump processes in the high-temperature limit comprise a phonon activation into a coincidence configuration, where the potentials of adjacent interstitial sites are equalized, and tunneling through the remaining barrier reduced by phonon interaction (Sect.2.3). The energy of the multiphonon process necessary to provide the appropriate adiabatic coincidence configuration, where tunneling occurs instantaneously, is isotope dependent – isotopes of larger mass need a higher activation in order to provide sufficiently large tunneling matrix elements. The local mode measurements, which probe the average vibrational potential, are not sensitive to occasional phonon fluctuation leading to jump processes and thus cannot distinguish between lattice-activated classical or quantum-jump processes. However, the quantum-mechanical treatment naturally explains the change of activation energies for H diffusion in Nb and Ta /2.19/, while classical lattice-activated diffusion fails to describe this phenomenon. Thus the outcome of the local mode measurements are consistent with the picture of H diffusion outlined in Sects.2.3 and 4.1.

4.3.2 H Band Modes

While the local vibrations of H in metals have been widely investigated, not much attention has been paid to H motions in the region of the acoustic host vibrations where the hydrogen was considered to move in phase with the host atoms. Only recent-

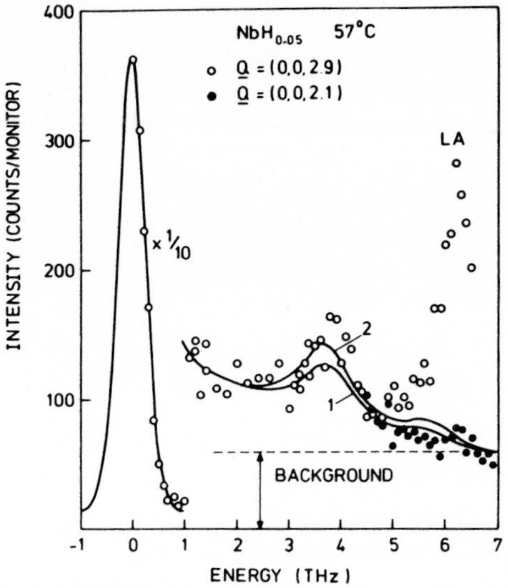

Fig.4.26. Spectrum for $NbH_{0.05}$ in the band-mode region; (Line 1) 1-parameter model; (Line 2) 6-parameter model. The peak LA origins from the host lattice phonons

ly theoretical as well as experimental investigations revealed resonant-like enhancements of the H amplitudes for certain lattice modes. LOTTNER et al. /4.82/ reported the observation of a strong peak at 15.7 meV $\hat{=}$ 3.8 THz in $NbH_{0.05}$ which is shown in Fig.4.26. It appears at the position of the peak in the density of states for the transverse acoustic modes in Nb. However, no comparable intensity enhancement is observed corresponding to the longitudinal cutoff which occurs at 25 meV. Thus, the underlying mechanism cannot simply be that of protons merely mirroring the acoustic phonons of the lattice. The authors propose an interpretation of this feature as a resonant-like mode of the protons within the acoustic vibrations of the lattice. The local density of state $H_i^H(\nu)$ for hydrogen at low frequencies ν was calculated for two models by standard Green function techniques /4.83/.

Model (1) is the simplest possible approach, and considers only a longitudinal coupling constant f between the H and its nearest Nb neighbors in analogy to (4.44), which describes reasonably well the local H vibrations. In this case the local spectrum can be given analytically

$$Z_i^H(\nu = \frac{\omega}{2\pi}) = 4M^H\omega \mathrm{Im}\{M^H\omega^2 - \varphi_i/[1-\hat{g}_i(\omega)f]\}^{-1} \quad , \tag{4.48}$$

where φ_i are elements of the dynamical matrix given by (4.44) and $\hat{g}_i(\omega)$ are appropriate combinations of elements of the lattice Green function for pure Nb. Since the Nb-H coupling strength is very large and M^H is small, the term $M^H\omega^2$ can be omitted for the band modes, and (4.48) reduces further to

$$Z_i^H(\nu) = 4M^H\omega \frac{1}{b_i} \mathrm{Im}\{|\hat{g}_i(\omega)|\} \quad , \tag{4.49}$$

where $1/b_i$ is a geometrical factor. Equation (4.49) does not depend any more on the coupling strength and is solely determined by geometry. Thus the H band modes mirror the local spectrum of the host phonon which may be slightly modified by the presence of H. Since $g_i(\omega)$ comprises combinations of elements of the lattice Green function, certain modes may appear more strongly than others. Model (2) was designed to account for the isotropic long-range displacement field in addition to the local vibrations, i.e., for the double force tensor P, and the change in the elastic constants due to H loading. For this purpose, transverse coupling of the H as well as longitudinal and transverse coupling changes between the four Nb neighbors were also considered. The corresponding 6 parameters were adjusted to the local mode frequencies, the dipole tensor P_{ij} /4.17/ and the changes of the elastic constants ΔC_{ijk} /4.84/. In Fig.4.27 the spectra in z direction obtained for the two models are compared with the respective host spectra. The spectra were divided by the corresponding masses in order to be proportional to the vibrational amplitudes of H and the metal atoms. Though only a small fraction of the H spectrum appears in the acoustic region (1.4% in z direction), the amplitudes are in the same order as those of the host. The most striking feature is a pronounced peak at 16 meV (1 THz ≙ 4.13 meV) which coincides with the (110) TA_1 zone boundary phonons. The H vibrates with enhanced amplitude in phase with the host atoms, as shown in the insert of Fig.4.27, in order to avoid a strain of the strong coupling constant f. Model (2) increases this effect even more.

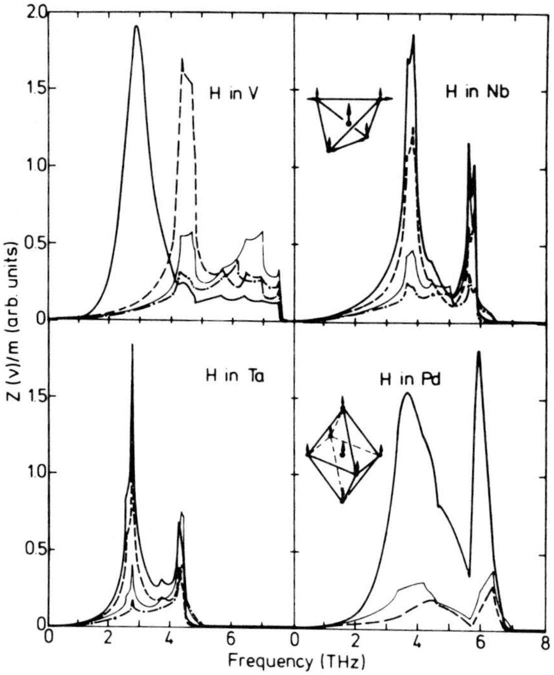

Fig.4.27. Band spectra for H. (———) $Z_Z^H(\nu)$ Model 2, (-----) $Z_Z^H(\nu)$ Model 1, (-·-·-) $Z^H(\nu)$ Model 2, (———) host spectrum. Inserts: schematic representation of z vibrations in bcc and fcc lattices

Due to the strong coupling of H and host vibrations in this frequency range, the isotope effect should be negligible. The solid lines in Fig.4.26 show the theoretical calculations in the context of the experimental findings. They were obtained considering damped harmonic oscillators for the hydrogen states in order to account for lifetime effects due to jump diffusion. The damping parameter was taken as the jump rate $1/\tau = 6\ D/d^2$ which was fitted in the quasielastic range. Normalization was performed between 1 and 3 THz. Both models are able to describe the observed peak around 16 meV. However, Model (2) is in more quantitative agreement with the experiment than Model (1).

Equivalent calculations have also been performed for V, Ta and Pd which all show similar resonant-like features. But besides some experimental hints like the observation of a band peak in Pd-Ag hydrides by CHOWDHURY /4.85/, systematic investigations of these materials are still missing.

The existence of resonant-like enhancements of the hydrogen amplitudes show that for certain modes the lattice exerts a large pull on the hydrogen which may induce or at least support diffusional jumps. In this connection the authors suggest a classical diffusion mechanism where the H is imagined as moving in the effective potential provided by the relaxed host atom positions. Assuming a smooth shape of the potential (e.g., sine-like) it is given by the static Green function G_{ij}^{HH} and the jump vector $\underline{s}$:

$$V_{eff}(x) = \frac{1}{2\pi^2} \sum_{ij} s_i G_{ij}^{HH}(0)^{-1} s_j \sin \frac{2\pi}{a} x \quad . \tag{4.50}$$

The activation energy becomes obviously

$$E_a = \frac{1}{\pi^2} \sum_{ij} s_i G_{ij}^{HH}(0)^{-1} s_j \quad . \tag{4.51}$$

Applying Model (2) activation energy values of 65, 73, and 37 meV were found for H diffusion in Nb, Ta and V.

Though agreement between the theoretical values and the experimental low-T data for H is rather good (Table 4.1), we consider it as fortuitous. In particular, it does not describe the large isotope effects observed at low temperatures where the activation energy for H diffusion changes while that for deuterium stays at the high-T value.

SUGIMOTO and FUKAI /4.86/ evaluated the H potential for the relaxed lattice configuation on the basis of Born-Mayer potentials whose parameters were chosen such as to describe the local H vibration and the isotropic dipole tensor. After solving the appropriate Schrödinger equation, they found an excited state which has large amplitudes near the neighboring pair of T sites and could be identified with a saddle-point configuration for over-barrier jumps. However, its energy of 167 meV is more than twice as high as the classical estimate by LOTTNER et al. /4.82/ and demonstrates the sensitivity of such estimates on the potential parameters as well as the importance of quantum-mechanical calculations.

Both calculations utilize static properties only and do not take advantage of the resonant-like enhancements of the H amplitudes, which may constitute experimental evidence for lattice-activated jump processes as proposed by EMIN /2.19/ (Sect.2.3). A theory which comprises the lattice dynamics as performed by LOTTNER et al. /4.82/ and the quantum-mechanical treatment of the H wave function in the fluctuating potential would be desirable.

4.4 Dynamics of Hydrogen Trapped at Impurity Atoms

Besides the effect of hydrogen trapping on hydrogen diffusion properties recently the dynamics of protons in the trapped state also came into focus: the observation of anelastic relaxation in Nb containing H *and* O or H *and* N and none in the presence of only one type of interstitial /4.41/ showed the trapping capability of the interstitial impurities and revealed the existence of local relaxation processes of the elastic dipole formed by the interstitial O-H or N-H pair. Measurements of their temperature dependence yielded information on the underlying jump mechanism /4.87-90/, while investigations of relaxational behavior as a function of crystal orientation indicated their point symmetry /4.88,42/. Finally, from the temperature-dependent relaxation strength, the pair binding energies can be evaluated. Recently also internal relaxation of H trapped at substitutional impurities like Ti in Nb has been reported /4.9/.

At low temperatures specific-heat /4.92,93/ and thermal conductivity /4.94/ experiments revealed a specific heat anomaly in Nb which appeared only in the presence of both N *and* H, and exhibits a pronounced isotope effect. This phenomenon was ascribed to H tunnel states in the neighborhood of the trapping impurity. Low-T anelasticity measurements support this picture /4.95/.

Very recently inelastic neutron scattering allowed a direct observation of the H tunneling transitions at low T /4.96/. At higher T, QNS measurements /4.97/ showed a local jump mechanism many orders of magnitude faster than that observed with internal friction at the same temperature. Finally, the observation of local H vibrations of trapped protons led to an unambiguous assignment of the H sites in the trapped state /4.43/.

4.4.1 Hydrogen Jump Processes Near a Trapping Impurity

Anelastic relaxation studies

Point-like impurities in a metal induce local lattice strains which in lowest order can be understood as a result of permanent elastic dipoles associated with the impurities. The dipole strength is measured by the double-force tensor P (2.1). If an external stress is applied on a crystal containing impurities, it may induce an additional time-dependent strain $\varepsilon_a(t)$ which originates from local reorientation of elastic dipoles (Snoek effect) or from inhomogeneous strains from the change of the impurity concentration profile in the sample (Gorsky effect). Here we are concerned with the Snoek effect.

The internal friction ϕ of a linear anelastic system may be written as /4.42/

$$\phi = \Delta \frac{\omega\tau}{1+\omega^2\tau^2} , \tag{4.52}$$

where ω is the measuring frequency and τ is the local relaxation time of the pertinent impurity. The relaxation strength Δ is proportional to the number of relaxing impurities, inversely proportional to temperature and related to the elastic properties of impurity and host /4.98,99/. Selection rules describing how an elastic dipole of a given symmetry couples to the various compliances have been discussed by NOWICK and HELLER /4.98/ and allow conclusions on the impurity symmetry, if measurements on single crystalline samples in different crystal directions are carried out.

Already early experiments revealed a low-temperature internal friction peak in Nb-H systems which was ascribed to H and D Snoek relaxations /4.89/, which would be expected for a nonisotropic double-force tensor. Later BAKER and BIRNBAUM /4.41/ showed that these peaks appear only in the presence of O or N *and* H. Consequently they were explained by a reorientational relaxation of the O-H or N-H pairs. For a temperature-independent concentration of relaxing species, the condition of maximum internal friction is reached according to (4.52) for $\omega\tau = 1$. Thus varying the temperature at different frequencies of the exciting elastic wave allows a determination of the temperature dependence of τ^{-1}. BAKER and BIRNBAUM found relaxation rates τ^{-1} which changed with impurity concentration. This concentration dependence was attributed to a possible superposition of different relation rates caused by N-H and O-H pairs (the O/N ratio varied with sample purity). More recently, these experiments were repeated and extended to better characterized samples using hydrogen as well as deuterium /4.87/. Single crystalline UHV purified Nb samples were doped with controlled amounts of O and H or D. The experiments were performed between 80 and 350 K in the frequency range 30 - 210 MHz. The longitudinal wave propagation direction was

(100). The most interesting feature of these measurements is the pronounced isotope effect in the H relaxation rate as well as in the O-H binding energy which was obtained from the T dependence of the relaxation strength [$E_a^H = 160 \pm 10$ meV, $E_a^D = 200 \pm 20$ meV, $E_b^H = 90 \pm 10$ meV, $E_b^D = 130 \pm 20$ meV – compare also with the μ^+ dissociation energy from a N trap of $E_a^{\mu+} = 48$ meV (Sect.3.5.2)].

Recently the internal friction measurements on O-H (N-H) pairs in Nb were extended to low temperatures and long relaxation rates /4.90,42/. Strain relaxation experiments allowed monitoring of jump rates down to 10^{-3} Hz. The experiments were performed on UHV purified polycrystalline Nb samples doped with controlled amounts of O or N. While around 80 K the relaxation frequencies of both pairs were about equal, toward lower T the N-H relaxation frequencies fall below those for the O-H pair. Figure 4.28 presents the observed O-H frequencies as a function of inverse temperature. Included are the earlier results obtained at higher T. Figure 4.28 clearly shows strong deviations from an Arrhenius behavior at low temperatures. The authors attributed this behavior to a proton transfer by incoherent tunneling. The data were fitted with Stoneham's closed expression for the small-polaron jump rate which combines the high- (2.6) and low- (2.9) temperature expressions /2.14/. Three parameters were fitted: the activation energy E_a, the tunneling matrix element J and, in order to account for a changed local phonon spectrum, the (local) Debye temperature Θ_D^ℓ. The result of the fit for O-H pairs is indicated by the dashed line in Fig.4.28, the evaluated parameters are listed in Table 4.4.

Considering the presumable multistep reorientation mechanism involving partial dissociation in a highly distorted lattice near the impurity, the application of the small-polaron theory in its simplest form appears to be doubtful. In particu-

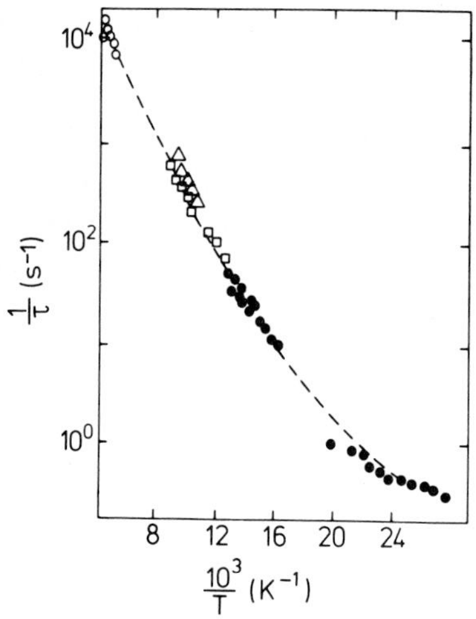

Fig.4.28. The logarithm of the H hopping frequency around an O impurity versus 1/T. (o) MATTAS /4.87/, (□) SCHILLER /4.88/, (•) CHEN /4.90/, (△) CANELLI /4.89/

Table 4.4. Tunneling parameters from a fit of internal friction results with small-polaron theory /4.42,87/

	E_a	θ_D^{ℓ}	J
N-H	222 meV	243 K	239 meV
O-H	207 meV	257 K	111 meV

lar, the values for the tunneling matrix element J seem unrealistically high in the light of what is known from μ^+ diffusion (Sect.3.5,6). In addition, small-polaron theory within the Condon approximation cannot account for the observed isotope effects, comprising a small change of the prefactor (O-H: $1.3 \cdot 10^{12}$ s^{-1}; O-D: $2.5 \cdot 10^{12}$ s^{-1}) /4.90/ and a considerable difference in the activation energies similar to the high-T results in Nb (Table 4.1). There, lattice-activated processes leading to adiabatic transfers were of great importance (Sect.4.1) and one could speculate about a similar situation for O-H(D) pairs.

Internal friction measurements for stress-wave propagation in the main symmetry directions of single crystalline NbO_xH_y samples were undertaken in order to gather information on the point symmetry of the impurity pair. Measuring at 1.37 and 2.35 KHz, SCHILLER and NIJMAN /4.88/ reported a relaxation peak in (111) direction and the absence of such a peak in (100) direction. They concluded that the O-H pair has its stable position parallel to (111) directions, while MATTAS and BIRNBAUM /4.87/ found a coupling to (100) stress waves in the MHz range indicating a (100) symmetry. Very recently, ZAPP and BIRNBAUM /4.42/ reinvestigated the orientational dependence of the O-H relaxations for all three main crystal directions in the MHz regime which corresponds to temperatures between 200 and 250 K. Relaxational peaks were found in all three symmetry directions. While in (100) and (111) directions the widths of the relaxation peaks indicated single relaxation processes, the broader peak in (110) direction suggested a composed relaxational behavior. Thus the (100) and (111) processes appear to be the independent relaxations. In order to account for the experimental results, two models were suggested; both assume H to occupy tunnel states comprising the tetrahedral and triangular sites on the face of a cube which are effectively centered at the octahedral sites. Such tunneling states were originally proposed by FLYNN and BIRNBAUM /4.100/ in order to describe low-T specific-heat anomalies and will be discussed in this context. The oxygen is known to occupy octahedral sites; the O-H pair configurations for the first three neighbors are displayed in Fig.4.29. Model (1) considers the second-nearest neighbor pairs as the most stable configuration, while Model (2) assumes first- and third-neighbor pairs. Both assumptions can account for the experimental results, while the presence of all three pair

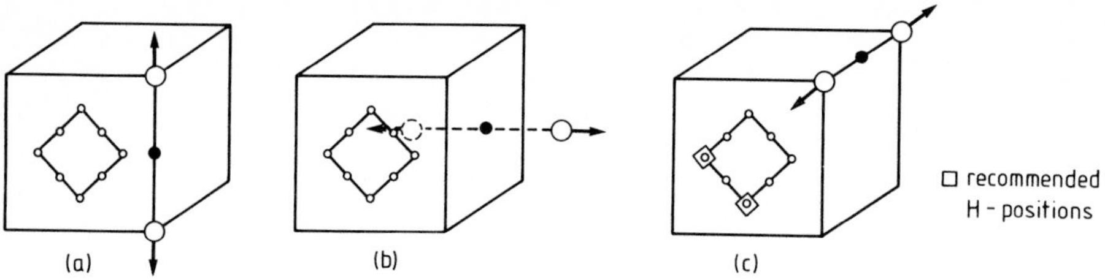

Fig.4.29a-c. O-H pair configurations in a unit cell of the Nb lattice. O atoms at the octahedral sites are shown by filled circles, the heavily displaced first Nb neighbors are indicated by large open circles: the attached arrows show the main direction of displacement. Finally the H tunnel rings are represented by small open circles; (a) first-, (b) second-, (c) third-neighbor configuration; (c) also indicates the most probable pair configuration suggested in /4.43/ (Sect.4.3.1)

configurations would lead to too large linewidths. Both cases lead to complicated reorientational paths requiring partial dissociation of the interstitial pairs.

The model assumption of tunnel-split proton states distributed over the four tetrahedral and four triangular sites on the face of the cube, however, is subject to considerable controversy. Within accuracy of a few percent, measurements of the proton form factor, which is the Fourier transform of the proton distribution for long times, revealed a point-like proton in the trapped state at 5 K /4.101/. Also at higher T no indications of a spread-out proton has been found but rather a fast local jump process has been detected /4.97/ which will be presented later on. Finally the localized vibrations of the hydrogen in an O-H pair configuration clearly reveal the tetragonal symmetry of a tetrahedral site /4.43/. Also theoretical arguments are against such a ring state containing H sites which are crystallographically inequivalent with respect to the O atom and therefore are split energetically. In the close neighborhood of the impurity those splittings can easily amount to some 10 meV which would suppress the tunneling process completely.

Very recently internal friction peaks have also been observed for Nb containing substitutional Ti impurities /4.91/. Low as well as higher temperature relaxation peaks occur in the same sample and have been attributed to the relaxation of TiH_x complexes at high T and to relaxation processes associated with tunneling states at low T. However, the sample was badly characterized and contained up to 1000 ppm O, N, or C; also the Ti concentration of 5% was too high for single relaxation centers. Future experiments on better defined samples should reveal further interesting information on hydrogen dynamics trapped at substitutional impurities in Nb.

QNS results

The self-correlation function of a proton exhibiting a spatially restricted motion does not decay to zero for infinite time but rather assumes finite values reflecting the finite probability to find the proton at a certain site for $t \to \infty$

$$G(\underline{r},t \to \infty) = G_{\infty}(\underline{r}) \quad . \tag{4.53}$$

The square of the Fourier transform of $G_{\infty}(\underline{r})$ is also called incoherent structure factor

$$F(\underline{q}) = \left| \int_{-\infty}^{+\infty} d^3r \, G_{\infty}(\underline{r}) \, e^{i\underline{q}\underline{r}} \right|^2 \quad . \tag{4.54}$$

It appears as an elastic contribution in the incoherent scattering law and bears information on the spatial extent of the proton motion. For purely vibrational motion, e.g., F(q) would be given by the Debye-Waller factor (4.7). For an arbitrary relaxational process which is restricted in space, the incoherent scattering law neglecting vibrational motion has the form

$$S^D_{inc}(\underline{q},\omega) = F(\underline{q})\delta(\omega) + \frac{1}{\pi}\sum_{i=1}^{\ell} \alpha_i(\underline{q}) \frac{\lambda_i}{\lambda_i^2+\omega^2} \quad , \tag{4.55}$$

where λ_i are the eigenvalues and $\alpha_i(q)$ are related to the eigenvectors of the corresponding system of rate equation (Sect.4.2.2). Further F(q) and the $\alpha_i(q)$ obey the sum rule

$$F(\underline{q}) + \sum_i \alpha_i(\underline{q}) = 1 \quad . \tag{4.56}$$

For a simple dumbbell configuration the orientationally averaged equation (4.55) assumes the form

$$S^D_{inc}(q,\omega) = \frac{1}{2}\left[1 + \frac{\sin q\ell}{q\ell}\right]\delta(\omega) + \frac{1}{2}\left[1 - \frac{\sin q\ell}{q\ell}\right]\frac{1}{\pi}\frac{1/\tau_\ell}{(1/\tau_\ell)^2+\omega^2} \quad , \tag{4.57}$$

where ℓ is the jump length and $1/\tau_\ell$ the local-jump rate. Thus, quasielastic neutron scattering on trapped protons exhibiting local jumps can reveal both the local relaxation rate *and* the spatial extent of the H motion. In the case of single crystals the crystallographic orientation of the jump process can also be obtained.

Recently QNS experiments on a polycrystalline $NbN_{0.004}H_{0.003}$ sample in the T range between 150 and 220 K have been performed /4.97/ applying high-resolution time of flight techniques. A typical experimental result is shown in Fig.4.30, where spectra obtained at q = 0.8 Å^{-1} and 1.8 Å^{-1} at 150 K are shown. While at q = 0.8 Å^{-1} the spectrum is completely elastic, at q = 1.8 Å^{-1} it consists of a superposition of two lines, a broad line (FWHM ≅ 100 μeV) containing 25% of the intensity and an elastic contribution. With decreasing q the quasielastic intensi-

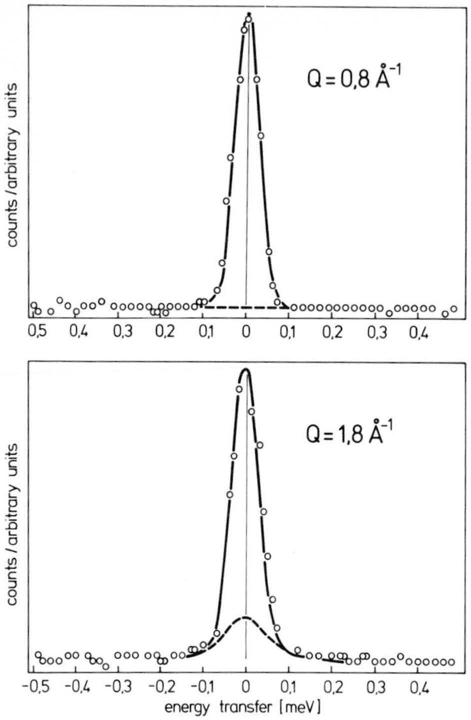

Fig.4.30. Spectra obtained from $NbN_{0.004}H_{0.003}$ at T = 150 K. The solid line represents a fit with a superposition of an elastic and a quasielastic contribution. The dashed line indicates the quasielastic fraction

ty drops while the linewidth stays practically constant. At a temperature where most of the H is trapped this result can be understood only as a signature of a local relaxation process of the trapped proton. Assuming a double well potential, a combined fit of the spectra with (4.57) revealed a jump length of 1 Å very close to the distance of two tetrahedral sites. The corresponding time scale is 4 orders of magnitude faster than the reorientation rate found by internal friction /4.42/. Its spatial extent of only 1 Å, however, is much smaller than the distances which have to be covered in the orientational relaxation processes.

This suggests an explanation in terms of fast local H jumps occurring between two neighboring tetrahedral sites near the impurity and an orientational relaxation of the dumbbell to another pair of H sites between which again fast local jump processes can take place. The QNS experiments revealed no sign of local tunneling states at 150 K. Also motion extending over the 4 tetrahedral and 4 triangular sites can be excluded. The temperature dependence of $1/\tau_\ell$ has not yet been established, but promises to reveal deeper insight into the mechanism of elementary H jumps at low temperatures.

4.4.2 Tunneling States of Trapped Hydrogen at Low Temperatures

Macroscopic measurements

Some years ago, SELLERS et al. /4.92/ reported the observation of a low-temperature specific-heat anomaly in Nb. The strong isotope effect made it clear that the observed feature was caused by the H or D atoms. BIRNBAUM and FLYNN /4.100/ ascribed this phenomenon to proton tunneling among tetrahedral and triangular interstices on the face of a cube as shown schematically in Figure 4.29 for the case of trapped protons. This model yields a H ground state split into 8 levels including two doublets. The degenerate multiplets may be split by random strains and cause a whole spectrum of tunneling states. Fitting the model to the specific heat data of Sellers et al. achieved good agreement with the experimental data.

Recently the relation between the specific-heat anomaly and the presence of N impurities in Nb has been investigated systematically /4.93/. The specific heat of UHV purified Nb, Nb doped with N or H and Nb doped with N and H or D was measured in the temperature region 0.07 - 1.5 K. Figure 4.31 shows the specific-heat data obtained together with the results by Sellers et al. They clearly demonstrate (i) that pure Nb as well as Nb doped either with H or with N exhibit essentially the same low-temperature behavior which is very close to the expected calculated curve for pure Nb; (ii) that samples containing both N and H or D show a large isotope-dependent specific-heat anomaly. Thus, the anomaly is related to the presence of both the N impurity and the H atom which at the low temperatures in question is trapped at the impurity. From the reported low residual resistivity ratio RRR = 100 it can be concluded that Sellers et al. already observed the same phenomenon. The relatively low size of the anomaly (the entropy increase due to the specific-heat anomaly is only 0.055 k and 0.087 k per trapped H and D, respectively) was attributed to stress-induced interaction effects among neighboring N-H pairs. On the basis of elastic continuum theory the average disturbance due to the neighboring defects was estimated to $\cong 1$ meV which is larger than the thermal energy of the anomaly. Their influence on the experimental result, therefore, is quite understandable.

POKER et al. /4.95/ reported low-temperature ultrasonic attenuation measurements on NbO_xH_y showing relaxation phenomena as displayed in Fig.4.32. While low-T relaxation was seen in $C' = (C_{11}-C_{12})/2$, none was observed in C_{44}. The frequency shift $\Delta f/f = \Delta c/c$ where c is the velocity of sound, is in the order of 20 ppm (f = 10 MHz). Between 2 and 3 K an increase in Δf is accompanied by a maximum in damping characterized by the decrement δ [δ corresponds to ϕ in (4.52)]. The strong decrease of Δf below 2 K has no counterpart in the damping. The authors interpret their results in terms of the tunneling model of BIRNBAUM and FLYNN /4.100/. Firstly, in order to understand the coupling to C', the third-nearest-neighbor configuration of Fig.4.29

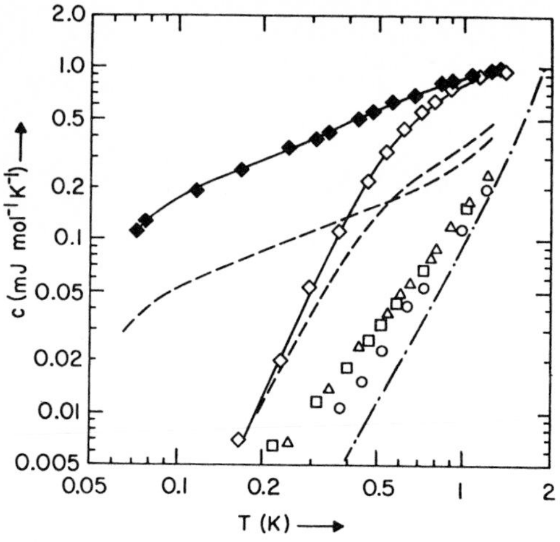

Fig.4.31. Specific heat of Nb samples partially doped with H, D and N in a log-log plot. (o) pure Nb, (□) $NbN_{0.003}$, (△) $NbH_{0.002}$, (◇) $NbN_{0.006}H_{0.002}$, (◆) $NbN_{0.003}D_{0.002}$. The solid lines are guidelines for the eye. The dashed lines represent the results of SELLERS et al. /4.92/. The dashed-pointed line gives the calculated specific heat for Nb

Fig.4.32. Decrement per unit concentration and frequency change versus temperature for H in NbO_xH_y obtained from internal friction

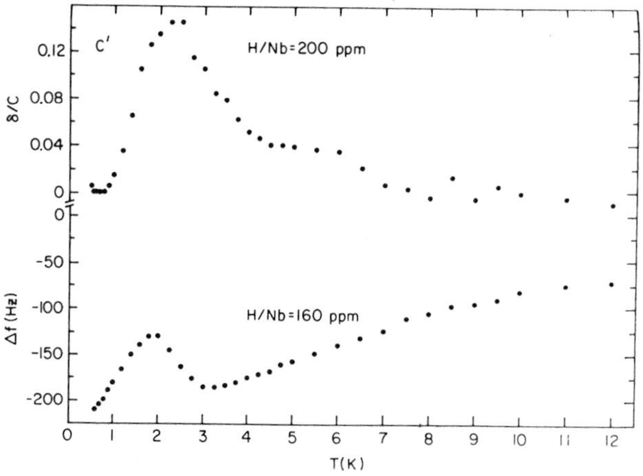

had to be invoked as the stable configuration. The observation of a coupling to C' is in agreement with earlier findings of SCHILLER and NIJMAN /4.88/ and implies that the (100) coupling reported by ZAPP and BIRNBAUM /4.42/ occurs only at higher temperatures, where a seemingly not so favorable 100 pair configuration exists. For the relaxation process an Orbach-type mechanism between the first excited states was proposed. It comprises an excitation to a higher level of the same symmetry and a subsequent downwards transition back to the other first excited state. At low temperatures the relaxation strength falls to zero when the excited states are no longer occupied and explains the decrease of Δf in Fig.4.32 below 2 K. Though the assumption of tunnel split ring states appears to fit the experimental data quite well, the same objections hold with respect to their physical soundness as put forward in Sect.4.4.1.

Inelastic neutron scattering results

Very recently inelastic neutron scattering experiments /4.96/ allowed a first direct spectroscopic observation of H tunneling in NbO_xH_y which has been interpreted in terms of tunneling in a disturbed double well potential as displayed schematically in Fig.4.33. Before we discuss the results, we briefly derive the scattering law for this situation. The Schrödinger equation for hydrogen motion in a double well potential can be solved exactly only for energy shifts $\varepsilon = 0$. For $\varepsilon \neq 0$ and excited states well above the tunnel split ground state, approximate solutions can be given /4.102/. Let J_{eff} be the tunneling matrix element between both sites [in order to include the polaron effect, we have to use J_{eff} (Sect.2.1,5)], then the ground state splitting ΔE is given by

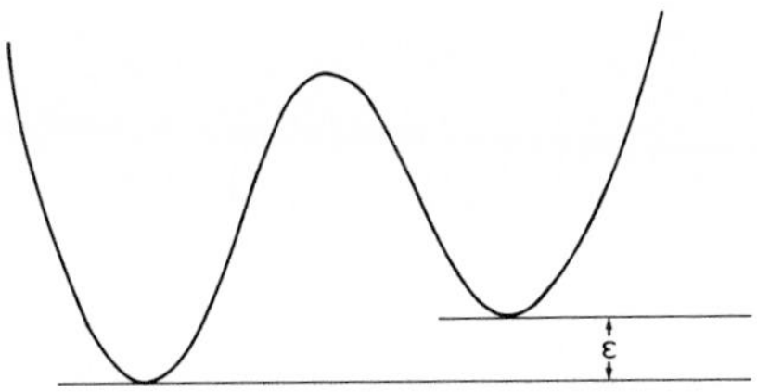

Fig.4.33. Schematic representation of a disturbed double well potential. The shift ε originates from the strain fields due to neighboring interstitial pairs

$$\Delta E = \sqrt{J_{eff}^2 + \varepsilon^2} \quad . \tag{4.58}$$

The corresponding proton wave functions are

$$|1(2)\rangle = \alpha|\ell(r)\rangle \pm \beta|r(\ell)\rangle \quad , \tag{4.59}$$

where $|\ell(r)\rangle$ are oscillator wave functions situated on the left- (right-) hand side. Further, α and β are given by

$$\alpha^2(\beta)^2 = \frac{\sqrt{J_{eff}^2 + \varepsilon^2} \pm \varepsilon}{2\sqrt{J_{eff}^2 + \varepsilon^2}} \quad . \tag{4.60}$$

The incoherent scattering function can be calculated from /4.103/

$$S_{inc}(\underline{q},\omega) = \sum_{i,j} p_i |\langle i|\exp(i\underline{qr})|f\rangle|^2 \delta[\omega-(E_i-E_f)] \quad , \tag{4.61}$$

where $|i\rangle$ and $|f\rangle$ denote the initial and final proton states, E_i and E_f are the corresponding energies and p_i are the thermal weight factors of the initial states. With (4.59,60), and after appropriate orientational averaging, we arrive at

$$S_{inc}(\underline{q},\omega) = \frac{1}{2}\left[\frac{J_{eff}^2+2\varepsilon^2}{J_{eff}^2+\varepsilon^2} + \frac{J_{eff}^2}{J_{eff}^2+\varepsilon^2}\,\frac{\sin q\ell}{q\ell}\right]\delta(\omega)$$

$$+ \frac{1}{2}\,\frac{J_{eff}^2}{J_{eff}^2+\varepsilon^2}\left[1-\frac{\sin q\ell}{q\ell}\right]\left[\frac{\delta(\omega+\Delta E)}{1+e^{-\beta\Delta E}} + \frac{e^{-\beta\Delta E}}{1+e^{-\beta\Delta E}}\,\delta(\omega-\Delta E)\right] \quad . \tag{4.62}$$

Three features are noteworthy: (i) for $\varepsilon = 0$ the elastic part has the same form as in (4.57) and can be understood as an interference pattern from a proton sitting at the same time in both potential minima; (ii) with increasing ε the elastic part increases, since an increasing part of the proton wave function is restricted to the deeper well; (iii) at $T = 0$, only energy-loss processes for the neutron are possible. With increasing temperature energy-gain processes appear, the intensity of which grows at the *expense* of the energy-loss processes, while for phonon scattering both intensities would grow.

The features obtained for the simple double well are also valid for more complicated structures. While the incoherent structure factor (4.54) is always the Fourier transform of the proton distribution for long times, for more complicated symmetries more than one pair of inelastic lines exist. For a distribution of disturbances $Z(\varepsilon)$, the scattering law has to be averaged with respect to it.

Figure 4.34 shows experimental spectra obtained on NbO_xH_y at 0.09 and 5 K, respectively. While at 0.09 K the spectrum is asymmetric, reflecting that only energy-loss processes of the neutrons are possible, at 5 K the data show an intensity

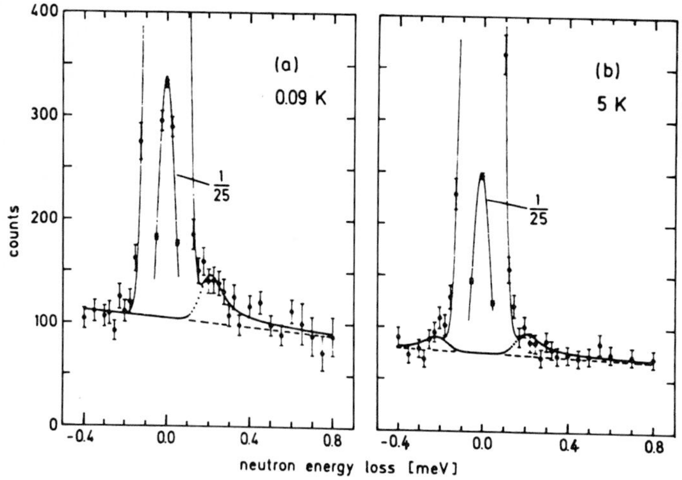

Fig.4.34a,b. Inelastic neutron spectra of $NbO_{0.013}H_{0.016}$ at (a) 0.09 K and (b) 5 K

increase on the energy gain side and a decrease on the energy loss side, as expected from (4.62) and in strong disagreement to phonon scattering. The data have been fitted quantitatively with (4.62) assuming a Lorentzian distribution of energy shifts ε created by the random strains due to the impurities /4.104/. According to STONEHAM, its width is given by a typical energy shift ε_0 between the two sites. The result of the fit is indicated by the solid line in Fig.4.34 and yielded an effective tunneling matrix element of J_{eff} = 0.19 meV in good agreement to the heat-capacity result of J = 0.18 meV /4.83/. Thereby only 1% of the total scattering intensity contributed to the inelastic line. This small intensity fraction demonstrates the importance of blocking effects firstly suggested to explain the specific-heat data. The neutron data do not allow an estimate of the structure factor and thus do not contain information on the spatial extent of the proton motion. Also the possible existence of further energetically higher tunneling states cannot be excluded. However, it is very likely that the relaxational process observed at 150 K is a reminiscence of the low-T tunneling process and that both processes occur within the same double well potential.

4.4.3 Local Vibrations of Hydrogen Trapped at Impurities

The degeneracy of the fundamental hydrogen vibrations is related to the point symmetry of the H interstitial position, e.g., for tetrahedral or octahedral sites in bcc metals two different fundamentals exist. While for the T site the upper frequency is degenerate, for the O site the reverse situation holds. For sites with lower symmetry like triangular interstices all three frequencies are different. This sensitivity of the H vibrations to their environment constitutes a powerful spectroscopic means to assign H interstitial positions and revealed the most reliable results concerning the position of hydrogen trapped by impurities. Furthermore, the positions of the vibrational frequencies yield information on the vibrational potential. Finally, the binding energy at a trap can be derived from the temperature-dependent intensity of spectral components associated with a trap.

Interstitial impurities

Figure 4.35 presents first spectra obtained from H trapped at N impurities in Nb at various temperatures /4.43/. The most striking feature of these results is the invariance of the peak position with temperature. The frequencies at 295 K where less than half of the H is trapped /2.39/ remain unchanged down to 10 K where all protons are in a trapped state. Similar results have also been obtained from $NbO_{0.011}H_{0.010}$ which demonstrate again that O and N are strong traps preventing hydride-phase separation, which would show itself with its easily recognizable intensity pattern (Fig.

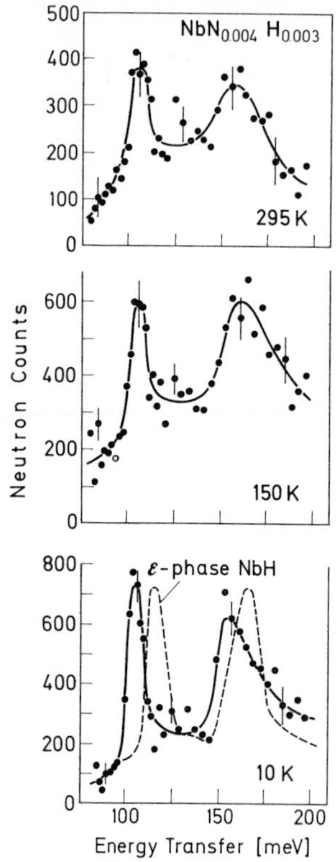

Fig.4.35. Inelastic spectra obtained from $NbH_{0.004}H_{0.003}$ at three temperatures. For comparison at 10 K the spectrum from the hydride phase (ε) is also included (-----)

4.24). The local vibrations of H trapped at O or N in Nb resemble closely the frequencies of the dilute α-NbH_x system and provide clear evidence that the sites occupied by trapped protons have a strong relationship to tetrahedral positions in a relatively undisturbed Nb environment. In particular, there is no evidence of any direct electronic force between the impurity and the H which would severely affect the tetragonal symmetry of the force field seen by the H. The experimental result of a tetrahedral site occupation is in disagreement with the tunnel split ring states put forward by BIRNBAUM and FLYNN /4.100/ and later again by ZAPP and BIRNBAUM /4.42/ (Sect.4.4.1) since no triangular site occupation is evident. It disproves also the idea of H atoms in so-called 4-T configuration, as suggested by FUKAI and SUGIMOTO /4.105/, where a H atom is delocalized over the 4 tetrahedral sites on the face of the cube and forms a common self-trapping distortion, leading effectively to the symmetry of an O site. Finally also the position between T and O sites as proposed on the basis of channeling experiments on TaN_xD_y /4.106/ can be excluded.

In order to assign a pair configuration MAGERL et al. /4.43/ took into account (i) the very short range Nb-H interaction potential /4.86/, which implies a large sensitivity of the H vibrations to the displacement of the host atoms and (ii) the requirement of at least two crystallographically equivalent nearest-neighbor H sites, in order to allow for the tunnel split ground state (Sect.4.4.2). Condition (i) excludes all H sites possessing a Nb nearest neighbor which is also nearest neighbor to the O impurity (Fig.4.29); (ii) suggests as the most probable configuration the nearest pair of tetrahedral sites in (111) direction from the impurity which also fulfills (i). Figure 4.29(c) shows that the suggested dumbbell configuration contains the two further away sites of the "ring state" for the third-neighbor O-H pair and accounts also for the symmetry requirements requested by the low-T internal friction measurements.

Both for H trapped at N as well as at O the linewidths of the vibrational peaks were found to be intrinsic. They are smaller than the corresponding widths in the dilute α phase but considerably broader than the peaks obtained from ordered hydrides. Since lifetime effects at low T are unlikely, three possible mechanisms, which could explain the large widths even at low T, were considered: (i) strain broadening which is expected to be rather effective in view of the short-range Nb-H interaction. It could also explain the larger linewidths observed for $NbO_{0.013}H_{0.012}$ than for $NbN_{0.004}H_{0.003}$, since the O concentration exceeded the N concentration by a factor of three; (ii) small disturbances from the tetrahedral symmetry which would lift the degeneracy of the upper mode and could provide a source of line broadening for this excitation; (iii) tunnel splitting of the excited states, which on the basis of the inelastic neutron scattering experiments on ground state tunnel splitting /4.96/ was estimated to contribute significantly to the line broadening of the first excited states (4 and 6 meV, respectively, for a harmonic potential). (iv) Acoustic optic multiphonon broadening which is also called recoil broadening /4.71/ could give rise to broadened wings around the inelastic peaks. In view of the resolution-determined local modes observed in Ta under similar conditions (Fig.4.25), this contribution can be discarded as a significant source for line broadening.

Thus combining all experimental evidence, we conclude that the complex relaxational behavior of H trapped at an O impurity in Nb can be visualized as follows:

(i) At low T, the hydrogen atoms occupy the pair of tetrahedral sites in (111) direction from the impurity as indicated in Fig.4.29.

(ii) The ground state is tunnel split with an effective tunneling matrix element of 0.19 meV.

(iii) Between the different (111) sites orientational relaxation can take place which gives rise to the internal friction peaks.

(iv) At higher temperatures, phonon interaction destroys the tunneling state and a fast relaxational process replaces it which was seen in the QNS studies at 150 K.

(v) Around 200 K O-H pairs in (100) directions are also occupied and give rise to the observed internal friction in this direction. This may be correlated with a strong broadening of the local hydrogen modes from the O-H pair at intermediate temperatures.

Substitutional impurities

Figure 4.36 shows first spectroscopic results on H trapping at the substitutional impurity Ti in Nb /4.44/. The peak positions together with the resolution corrected linewidths are displayed in Table 4.5. While position and linewidth of the lower fundamental change only weakly with temperature, the higher excitation which is only vaguely defined at room temperature sharpens considerably with decreasing T and shifts toward lower frequencies. At 10 K both frequencies are distinctly different from the ordered hydride phase vibrations which occur at 116 and 167 meV. Thus, Ti provides a strong trap for H in Nb which prevents phase separation into a hydride phase as indicated also by internal friction measurements /4.91/. Again the vibrational frequencies immediately show that the protons trapped at Ti impurities occupy sites with a symmetry very close to that of a tetrahedral site in Nb. As in the case of interstitial impurities, the linewidths are enhanced compared to those in the hydride phases and similar arguments can be used for their explanation. Other than the interstitial impurities, Ti appears to affect also the peak position of the upper vibration which is shifted by more than 10 meV from the α phase position.

Figure 4.37 presents the results of equivalent measurements taken on $NbV_{0.008}H_{0.005}$ /4.43/ for a sequence of temperatures. At room temperature, NbV_xH_y shows very broad

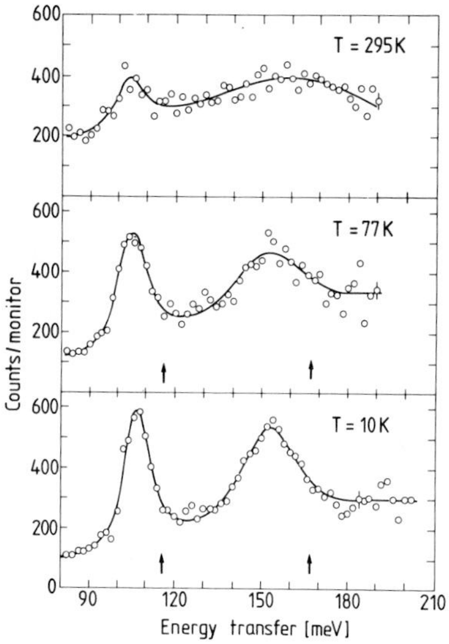

Fig.4.36. Inelastic spectra obtained from $NbTi_{0.01}H_{0.009}$ at three temperatures. The arrows indicate the peak positions in the hydride phase

Table 4.5. Vibrational frequencies of H in $NbTi_{0.01}H_{0.009}$ /4.107/

Temperature [K]	$\hbar\omega_1$ [meV]	FWHM [meV]	$\hbar\omega_2$ [meV]	FWHM [meV]
295	104 ± 1	8 ± 3	164 ± 6	-
77	105 ± 0.5	9 ± 1	153 ± 2	29 ± 6
10	107 ± 0.5	7 ± 1	153 ± 1	20 ± 2

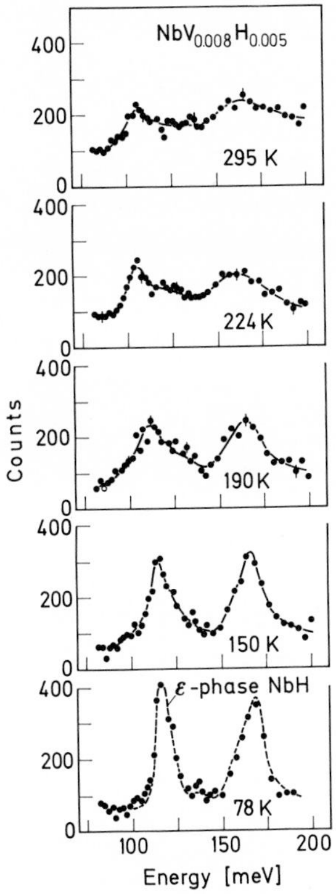

Fig.4.37. Inelastic neutron spectra obtained from $NbV_{0.008}H_{0.005}$ at various temperatures. The dashed line at 78 K represents the hydride phase result

density-of-state peaks which sharpen up somewhat as the temperature is reduced. Simultaneously they shift toward higher energies. At 150 K the spectrum appears to be composed of a broad and a narrow peak characteristic for a two-phase situation. At 78 K the two excitations are centered at the hydride-phase positions and the linewidths are greatly reduced.

Contrary to Ti which keeps the H trapped even at low temperatures, V does not attract H strongly enough to prevent phase separation. However, it does appear to shift the phase boundary for precipitation to lower temperatures. While at 190 K, where for pure α-NbH_x the solubility limit for H is x = 0.003 /4.39/, no evidence for hydride formation is found in $NbV_{0.008}H_{0.005}$, at 150 K the two-component lower peak indicates that about 50% of the H is already precipitated. Thus, V appears to shift the α-β phase boundary to higher concentrations without preventing the eventual precipitation of the hydride phase at low temperatures.

5. Outlook and Conclusion

This review has shown that diffusion studies on light interstitials in metals are a very vivid area of research and that there is great interest in the fundamental transport mechanisms. The transport phenomena of light interstitials including the muon are situated in a mid position between bandlike transport as performed by electrons in metals and classical over-barrier diffusion as exhibited by heavy atoms, e.g., in superionic conductors or during self-diffusion in metals. Diffusion studies on light interstitials appear to cover nearly the whole spectrum of transport processes starting from coherent tunneling of muons at low temperatures and reaching to nearly classical diffusion of H in Cu or Ni. This large variety of phenomena can be accessed by changing temperature and isotope, only. While the low-T transport of muons bears similarities to narrow-band electronic conductors /5.1,2/ and may also have relations to electronic conductivity in heavily disordered metals /5.3,4/, thermal-activated diffusion has its counterpart in electronic conduction in semiconductors and insulators /2.1,2/.

In particular the development of muon spin rotation provided a richness of new and unexpected results. As a consequence of muon implantation low and ultra low temperatures became accessible for diffusion experiments. At the same time the light mass of the muon caused quantum transport phenomena to become more prominent. Careful measurements on Al, which can be regarded as a prototype material like Pd in the case of H, led to the determination of the muon diffusion coefficient over 4 orders of magnitude in temperature, which to the author's knowledge is by far the largest range of any solid-state diffusion study. The essential results of this study are: (i) Clear experimental evidence for both incoherent hopping above 3 K and coherent muon motion below 3 K was established. (ii) The current small-polaron theory can neither account for the observed weak temperature dependencies of the diffusion coefficient, nor predict the low transition temperature of 3 K. Since Al is the only material investigated systematically at low T, results on other materials would be highly desirable in order to judge whether the observed transport phenomena are of universal nature.

Experiments on Nb gave rise to the conclusion that muons do not self-trap always immediately after implantation but rather may stay in a metastable propagating state. Self-trapping occurs by thermal fluctuations and can be induced by impurities. These experiments may have opened the door to investigations of the kinetics of small-polaron formation, and further interesting results are anticipated.

A drawback of most of the μ^+ diffusion studies up to now is that the results have been obtained indirectly through diffusion-controlled trapping since intrinsic muon diffusion is too fast to be detected directly. Therefore macroscopic diffusion experiments on small particles or thin foils which would allow a direct determination of the diffusion coefficient would be highly desirable in order to verify the outcome of the indirect measurements.

After having reached a premature understanding of the basic muon behavior in metals, it is of consequence to consider potential muon applications in metal physics. In particular, its striking sensitivity toward impurities and defects makes it in analogy to the positron /5.5/ a promising tool for the investigation of lattice imperfections /3.56/. Microscopic information on defect and impurity properties can be obtained at extremely low concentrations. Performing field- and orientation-dependent measurements on single crystals, the structure of the muon traps can be identified and, e.g., recovery from radiation damage can be studied in great detail.

For hydrogen in metals the course of research took a reversed path. The technical implications like H embrittlement, isotope separation, H storage, etc., requested knowledge on the elementary processes, which are studied with a large variety of macroscopic and microscopic methods. Diffusion results on the standard fcc and bcc metals are well established and reach down to about 150 K. Other than for muons the existing anomalies can be interpreted in terms of small-polaron theory beyond the Condon approximation. For the future, in particular measurements at low temperatures on all isotopes would be desirable, in order to clarify whether the concepts applied today are pertinent.

For a study of the microscopic motional properties of H in metals, neutron scattering has proved itself as the method with the greatest potential, able to explore the space *and* time evolution of the diffusive process. Up to now most experiments were directed toward an investigation of the diffusional properties of single H atoms in typical fcc and bcc materials. Thereby only the "classical" experiments on Pd at low H concentrations revealed simple next-neighbor jumps. At higher temperatures correlated sequences of H jumps are present in bcc metals whose quantitative explanation is still not worked out. Now a tendency toward more complex systems is observable: experiments on H diffusion in the presence of trapping impurities led to a microscopic observation of trapping and escape processes from nitrogen atoms in Nb. Further experiments, e.g., on systems containing substitutional impurities, appear to be attractive. The investigation of space-time correlations in concentrated metal-hydrogen systems is just at the beginning. In particular, experiments on H sublattices of low coordination numbers are desirable. Increased future efforts are expected on technical H storage materials. There the diffusional process on an atomic scale is exceptionally intricate and combines aspects of trapping and concentration effects. First results on $Ti_{1.2}Mn_{1.8}H_3$ are encouraging and certainly will stimulate further experiments.

The investigation of the local-jump processes on trapped protons caused a considerable extension of the temperature range open for diffusion measurements on H in metals and reached an overlap with the domain of muon diffusion. Local diffusion can be followed down to low temperatures, and measurements in the low-T regimes of the small-polaron theory are possible. This includes the investigation of tunneling states and their coupling to the phonons of the host. Contrary to tunneling in molecular crystals, which is stabilized by the small spin lattice relaxation rate /5.6/, the full impact of the host phonons can be studied. In particular the transition from a coherent tunneling state to local hopping may be observable in the future.

The investigation of the relaxation between H vibrations and H diffusion may help to identify further aspects to the diffusional properties of light interstitials. A detailed experimental determination of the H vibrations yields important information on the H-host interaction, the knowledge of which is required for theoretical calculations on diffusion. The observation of resonant-like band modes may be associated with lattice-activated diffusion processes, and an atomistic identification of the related phonon modes is at hand.

Summarizing, the large body of experimental data demonstrates impressively that neutron scattering on H in metals and μSR are complementary in their results. μSR yields information on elementary diffusive processes at low T and infinite dilution. Its application as a diagnostical method in metal physics is just coming into focus. Neutron scattering on H in metals on the other hand is a well-established method and provides extensive information on the space-time correlations of protonic motion including vibrational properties. We expect a further mutual stimulation of muon and hydrogen diffusion studies and anticipate considerable theoretical efforts in order to provide conclusive interpretations.

A1. Appendix

Considering correlated averages over the fluctuations which lead to a reduction of the potential barrier and over fluctuations which equilibrate the small polaron levels, TEICHLER /2.20,21/ derived explicit expressions for the small-polaron jump rate. Thereby he assumed that the phonons enter the transfer matrix element $J_{pp'}$ only in combinations that characterize the height of the potential barrier U above the energy of the incident particle. For the jump rate above $\theta_D/2$ he found

$$\Gamma_{pp'} = \frac{1}{\hbar} \frac{\pi}{\sqrt{4E_a kT}} \langle |J_{pp'}|^2 \rangle \exp(-E_{pp'}/kT) \tag{A1.1}$$

$$E_{pp'} = (E_{p'} - E_p + 4E_a)^2/16E_a \quad , \tag{A1.2}$$

where E_p and $E_{p'}$ denote the energy levels (without phonons) of the interstitial-host metal system relaxed around the initial and final positions, respectively. The averaged tunneling matrix element is given by

$$\langle |J_{pp'}|^2 \rangle = \int_{-\infty}^{+\infty} d\varepsilon J_0(\varepsilon)^2 \exp[-(\varepsilon - \Delta U_p)^2/2\sigma^2]/(\sqrt{2\pi}\sigma) \quad , \tag{A1.3}$$

where ΔU_p is the barrier height above the interstitial level in position p and σ^2 measures the barrier height fluctuations due to phonons. For high temperatures we have

$$\sigma^2 \cong 8\, kT E_c \quad , \tag{A1.4}$$

where E_c is an energy characteristic for the magnitude of the barrier fluctuations.

In a one-dimensional model a further evaluation of (A1.1,3) is possible. Assuming a parabolic potential barrier around the saddle point $U(x) = U_s - kx^2/2$, the transmission coefficient $J_0^2(\varepsilon)$ is given by /2.22/

$$J_0^2(\varepsilon) = J_\infty^2/[1 + \exp(\varepsilon/\varepsilon_1)] \quad , \tag{A1.5}$$

with $\varepsilon_1 = (\hbar/2\pi)\sqrt{k/m}$. For high temperatures ($\sigma \gg \varepsilon_1$) $J_0^2(\varepsilon)$ can be approximated by a step function and (A1.3) yields an activated behavior for J

$$\langle |J_{pp'}|^2 \rangle = J_\infty^2 \frac{\sqrt{4kT\, E_c/\pi}}{\Delta U_p} \exp\left(-\frac{\Delta U_p^2}{16\, kT\, E_c}\right) . \tag{A1.6}$$

Considering only transitions between energetically equivalent levels and inserting (A1.6) into (A1.1), the Flynn-Stoneham result for lattice-activated processes (2.12) is retained.

To estimate the behavior of J at intermediate temperatures as a further simplification, the characteristic phonon frequencies were modeled by an Einstein spectrum of the frequency ω_E. Figure A1.1 presents results of such a calculation as a function of temperature and correlation q between fluctuations of the barrier height and fluctuations of the interstitial levels (q = 0: complete correlation, q = 1: no correlation). At low T tunneling transitions dominate, whereas at higher T the lattice-activated processes take over. The temperature dependence itself depends strongly on the degree of correlation between the different fluctuations, which are determined by the geometry of the lattice and the interaction potential between the H isotope and the host. For large correlations, distinct transitions from nonactivated to activated behavior appear to be possible.

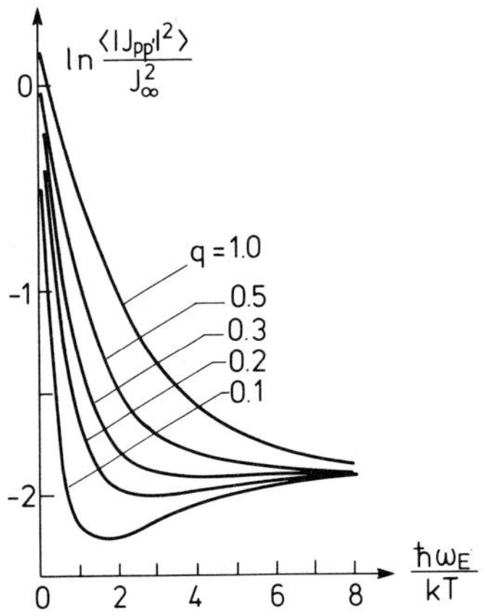

Fig.A1.1. Semilogarithmic plot of $\langle |J_{pp'}|^2 \rangle$ (A1.3) as a function of $\hbar\omega_E/kT$ /2.20/

A2. Appendix

The question whether the fluctuational reduction of the tunneling barrier by one-phonon processes may yield qualitatively new results has recently been discussed by TEICHLER /2.26/. He evaluated a transition rate which reads in the Debye approximation for $kT \gg \Delta E$

$$\Gamma^{1Ph} = \frac{3\,kT}{4h^2 \rho c^3 (2\pi)^3} \left| \langle\langle \sum_\alpha \frac{\partial}{\partial u_\alpha} J \rangle\rangle \right|^2 e^{-2S(T)} \quad , \qquad (A2.1)$$

where $J_{eff} = J \exp[-2S(T)]$ and J is assumed to depend linearly on the atomic displacements u_α. The expression within the brackets is the low-T equivalent of (A1.1) describing now one-phonon transitions. The important difference between the one-phonon rate beyond the Condon approximation and the rate within the Condon approximation (2.15) lies in the fact that (A2.1) does not depend explicitly on the disturbances ΔE. Teichler argues that $\langle\langle \partial J/\partial u_\alpha \rangle\rangle$ is different from zero for coincidence configurations which do not have point-inversion symmetry. While for bcc metals the symmetry requirements are fulfilled easily, in fcc metals inversion centers for jumps between tetrahedral and octahedral sites exist. He concludes that contributions can occur only if the effective coincidence situations correspond to atomic arrangements with broken symmetry. For $\langle\langle \partial J/\partial u_\alpha \rangle\rangle = 0$ two-phonon processes are the leading contribution which may change the prefactor in the jump rate (2.9) considerably but will not change its temperature dependence.

A3. Appendix

For combined electric and magnetic interaction, the Hamiltonian for the muon spin precession is the sum of the magnetic dipole interaction

$$H_{mag} = -\hat{\mu}_\mu \cdot B_0 \tag{A3.1}$$

and the electric quadrupole interaction

$$H_{el} = \frac{4}{5}\pi \, \underline{\underline{T}}\,\underline{\underline{V}} \quad , \tag{A3.2}$$

where $\underline{\underline{T}}$ and $\underline{\underline{V}}$ are the tensors of the electric quadrupole moment and the electric field gradient respectively. With the help of some spherical tensor algebra /3.17/ H_{el} can be transformed into the basis of $\hat{I}$. There, the (2I+1)-dimensional matrix of the combined electric and magnetic interaction has the following nonvanishing matrix elements:

$$H_{m,m} = -\hbar\omega_B m + \frac{1}{2}\hbar\omega_E (3\cos^2\beta - 1)(3m^2 - I(I+1))$$

$$H_{m,m-1} = -\frac{3}{2}\hbar\omega_E \cos\beta \sin\beta \, (1-2m)((I-m+1)(I+m))^{1/2}$$

$$H_{m,m-2} = \frac{3}{4}\hbar\omega_E \sin^2\beta((I+m-1)(I+m)(I-m+1)(I-m+2))^{1/2} \quad , \tag{A3.3}$$

where β is the angle between the directions of the electric field gradient and the magnetic field, ω_B is the Larmor frequency and ω_E is an electric interaction frequency

$$\hbar\omega_E = \frac{1}{2}\frac{eQq}{2I(I-1)} \quad , \tag{A3.4}$$

where Q is the value of the quadrupole moment and q that of the EFG. Since H is no longer diagonal with respect to I_z, also the transverse spin component I_x contributes to the local field ΔB_z at the muon site:

$$\Delta B_z = \gamma_i\hbar <I_z>^{stat} \cdot \frac{3\cos^2\vartheta-1}{r^3} + \gamma_I\hbar <I_x>^{stat} \cdot \frac{3\sin\vartheta\cos\vartheta}{r^3} \quad , \tag{A3.5}$$

$<I_x>^{stat}$ and $<I_z>^{stat}$, thereby, are the average static components of I_x and I_z. Assuming a $1/r^3$ dependence for the EFG, HARTMANN /3.16/ carried out the lattice sums over 70 nuclei around the interstitial positions, but the dominant contribution comes from the next neighbors. Theoretical field dependencies are shown in connection with experimental data (Sect.3.4)

A4. Appendix

For the case of coherent diffusion the theory of muon depolarization is only partly worked out. In particular a generalization of the correlation time τ_c, which for extended wave packets is not simply a mean residence time at a certain site, is required. For this purpose it is advantageous to express the polarization P(t) in terms of the time-dependent correlation function of the magnetic field acting on the muon. Under the assumption of negligible T_1 processes, McMULLEN and ZAREMBA /3.34/ showed that P(t) is determined by the self-correlation function $<\hat{n}(r,t)\hat{n}(r',t')>$ of the muon, where $\hat{n}(r,t)$ is the muon density operator. For sufficiently long times the decay rate of P(t) is governed by a correlation time

$$\tau_c = \int_0^\infty d(t-t') <n(r,t)n(r,t')> \tag{A4.1}$$

which for jump diffusion reduces to the mean residence time $\tau_c = \tau$. For coherent diffusion in the transition region, where the extension of the wave packets is still small, τ_c should be the effective time necessary for a transfer to a neighboring site

$$\tau_c \cong a^2/D_{coh} \quad . \tag{A4.2}$$

For long mean-free path $\ell >> a$ the theory has not yet been worked out. In this region only the case of ferromagnetic materials has been treated by FUJII and UEMURA /2.13,3.35/, but their results are specific for ordered magnetic structures.

Acknowledgement. The author thanks Prof. Dr. T. Springer and Drs. K.W. Kehr and R. Hempelmann for valuable comments on the manuscript and for numerous clarifying discussions. In addition, contacts with many members of the international μSR and H metals community helped to shape the picture of light interstitial transport phenomena presented here.

References

2.1 T. Holstein: Ann. Phys. (N.Y.) *8*, 325 (1959)
2.2 T. Holstein: Ann. Phys. (N.Y.) *8*, 343 (1959)
2.3 J. Yamashita, T. Kurosawa: J. Phys. Chem. Sol. *5*, 34 (1958)
2.4 C.P. Flynn, A.M. Stoneham: Phys. Rev. B *1*, 3966 (1970)
2.5 see, e.g., G. Leibfried, N. Breuer: *Point Defects in Metals I*, in Springer Tracts in Modern Physics Vol. 81 (Springer, Berlin, Heidelberg, New York, 1978)
2.6 H. Peisl: in *Hydrogen in Metals I*, Topics in Applied Physics Vol. 28 (Springer Berlin, Heidelberg, New York, 1978) pp. 53
2.7 H. Horner, H. Wagner: J. Physics C *7*, 3305 (1974)
2.8 G.H. Vineyard: J. Phys. Chem. Sol. *3*, 121 (1957)
2.9 V.N. Pavlovich, V.N. Rudko: phys. stat. sol. (b) *88*, 407 (1978)
2.10 M. Wagner: phys. stat. sol. (b) *88*, 517 (1978)
2.11 D.L. Tonks, B.G. Dick: Phys. Rev. B *19*, 1136 (1979)
2.12 A.M. Stoneham: in *Exotic Atoms 79*, ed. by K. Crowe, J. Duclos, G. Fiorentini, G. Torelli Ettore Mayorana, International Science Series Vol. 4, (Plenum Press, New York, 1980)
2.13 S. Fujii: J. Phys. Soc. Japan *46*, 1833 (1979)
2.14 A.M. Stoneham: J. Physics F *2*, 417 (1972)
2.15 K.W. Kehr: Jül-Report JÜL-1211, Kernforschungsanlage Jülich (1975)
2.16 H. Teichler: Phys. Lett. *64A*, 78 (1977)
2.17 H. Teichler: Phys. Lett. *67A*, 313 (1978)
2.18 K.W. Kehr: in ref. 2.6, pp. 197
2.19 D. Emin, M.I. Baskes, W.D. Wilson: Phys. Rev. Lett. *42*, 791 (1979)
2.20 H. Teichler: in ref. 2.12, pp. 283
2.21 H. Teichler: phys. stat. sol. (b) *104*, 239 (1981)
2.22 L.D. Landau, E.M. Lifschitz: *Quantenmechanik* (Akademie Verlag, Berlin, 1966)
2.23 J.A. Sussmann: Phys. Kondens. Materie *2*, 146 (1964)
2.24 R.J. Rollefson: Phys. Rev. B *5*, 3235 (1972)
2.25 H. Teichler, A. Seeger: Phys. Lett. *82A*, 91 (1981)
2.26 H. Teichler: Hyperfine Int. *8*, 505 (1981)
2.27 K.W. Kehr, D. Richter, J.M. Welter, O. Hartmann, E. Karlsson, L.O. Norlin, T.O. Niinikoski, A. Yaouanc: Phys. Rev. B *26*, 567 (1982)
2.28 P.W. Anderson: Phys. Rev. *109*, 1492 (1958)
2.29 Yu. Kagan, M.J. Klinger: J. Phys. C *7*, 2791 (1974)
2.30 J.D. Eshelby: *The Continuum Theory of Lattice Defects*, in Solid State Physics, Vol. 3 (Academic Press, New York, 1956)
2.31 B. Baranowski, S. Majchrzak, T.B. Flanagan: J. Phys. F *1*, 258 (1971)
2.32 W.B. Pearson: *Lattice Spacings and Structure of Metals and Alloys* (Pergamon, Oxford, 1958)
2.33 K.W. Kehr: Proc. Second JIM Int. Symposium Hydrogen in Metals, Minakami, Nov. 26-29 (1979); Suppl. Trans Jpn. Inst. Met. *21*, 181 (1980)

2.34 Yu. Kagan, M.J. Klinger: Sov. Phys. JETP *43*, 132 (1976)
2.35 A.F. Andreev, I.M. Lifshitz: Sov. Phys. JETP *29*, 1107 (1969)
2.36 J. Jäckle, K.W. Kehr: J. Phys. F, in print
2.37 T.McMullen, B. Bergersen: Solid State Comm. *28*, 31 (1978)
2.38 K.G. Petzinger: unpublished
2.39 D. Richter, T. Springer: Phys. Rev. B *18*, 126 (1978)
2.40 K.W. Kehr, D. Richter: Solid State Comm. *20*, 477 (1976)
2.41 K.W. Kehr, D. Richter, R.H. Swendsen: J. Phys. F *8*, 433 (1978)
2.42 T.R. Waite: Phys. Rev. *107*, 463 (1957)
2.43 K. Schroeder: in *Point Defects in Metals II*, in Springer Tracts in Modern Physics, Vol. 87 (Springer, Berlin, Heidelberg, New York, 1980)
2.44 C.H. Hodge: J. Phys. F *4*, L230 (1974)
2.45 I.K. Mackenzie, T.L. Koo, A.B. McDonald, B.T.A. McKee: Phys. Rev. Lett. *19*, 946 (1967)
2.46 A.M. Browne, A.M. Stoneham: AERE preprint TP. 880, Harwell (1980) and J. Phys. C *15*, 2709 (1982)
2.47 D. Emin: Hyperfine Int. *8*, 515 (1981)
2.48 C.H. Leung, T. McMullen, M.J. Scott: J. Phys. F *6*, 1063 (1976)
2.49 C.H. Hodge, H. Trinkhaus: Solid State Comm. *18*, 857 (1976)
2.50 D. Emin: Adv. Phys. *22*, 57 (1973)
2.51 D. Emin, T. Holstein: Phys. Rev. Lett. *36*, 323 (1976)
2.52 D. Emin: in *Muon Spin Rotation*, Proceedings of the Second International Topical Meeting on Muon Spin Rotation, Vancouver B.C., Canada (August 11-15, 1980), p. 522 (North Holland, Amsterdam, New York, Oxford, 1981)
3.1 C.D. Anderson, S.H. Neddermeyer: Phys. Rev. *50*, 263 (1936); *51*, 884 (1937); *54*, 88 (1938)
3.2 J.C. Striet, E.C. Stevenson: Phys. Rev. *51*, 1005 (1937)
3.3 H. Yukawa: Proc. Phys. Math. Soc. Japan *17*, 48 (1935)
3.4 R.L. Garwin, L.M. Ledermann, M. Weinrich: Phys. Rev. *105*, 1415 (1957)
3.5 For an overview see: *Muon Spin Rotation*, Proceedings of the First International Topical Meeting on Muon Spin Rotation, Rohrschach. Switzerland (September 4-7, 1978) (North Holland, Amsterdam, New York, Oxford, 1979); and
3.6 *Muon Spin Rotation II*, Proceedings of the Second International Topical Meeting on Muon Spin Rotation, Vancouver B.C., Canada (August 11-15, 1980) (North Holland, Amserdam, New York, Oxford, 1981)
3.7 see, e.g., C.S. Wu: *The Neutrino*, in Theoretical Physics in the Twentieth Century, ed. by M. Fierz and V.F. Weisskopf (Interscience, New York 1960)
3.8 A. Schenck: in *Nuclear and Particle Physics at Intermediate Energies*, ed. by J.B. Warren (Plenum, New York, 1976)
3.9 D.K. Bryce: Phys. Lett. *66A*, 53 (1978)
3.10 see, e.g., A. Abragam: *Nuclear Magnetism* (Oxford Univ. Press, Oxford, 1961)
3.11 R. Kubo, T. Toyabe: in *Magnetic Resonance and Relaxation*, ed. by R. Blinc (North Holland, Amsterdam, 1967)
3.12 T. Yamazaki: Hyperfine Int. *6*, 115 (1979)
3.13 K.G. Petzinger: Phys. Lett. *75A*, 225 (1980)
3.14 J.H. van Vleck: Phys. Rev. *74*, 1168 (1948)
3.15 O. Hartmann, E. Karlsson, K. Pernestal, M. Borghini, T.O. Niinikoski, L.O. Norlin: Phys. Lett. *61A*, 141 (1977)
3.16 O. Hartmann: Phys. Rev. Lett. *39*, 832 (1977)
3.17 E. Matthias, W. Schneider, R.M. Steffen: Phys. Rev. *125*, 261 (1962)
3.18 M. Camani, F.N. Gygax, W. Ruegg, A. Schenck. H. Schilling: Phys. Rev. Lett. *39*, 836 (1977)
3.19 O. Hartmann, E. Karlsson, L.O. Norlin, D. Richter, T.O. Niinikoski: Phys. Rev. Lett. *41*, 1055 (1978)
3.20 J.P. Bugeat, A.C. Chami, E. Ligeon: Phys. Lett. *58A*, 127 (1976)
3.21 C. Berthier, M. Minier: J. Phys. F *7*, 515 (1977)
3.22 P. Jena, S.G. Das, K.S. Singwi: Phys. Rev. Lett. *40*, 264 (1978)
3.23 P. Hohenberg, W. Kohn: Phys. Rev. *136B*, 864 (1964) and W. Kohn, L.J. Sham: Phys. Rev. *140A*, 1133 (1964)

3.24 J. DeLauney: in Solid State Physics 4, Vol. 2, ed. by F. Seitz, D. Turnbull (Academic Press, New York, 1956)
3.25 J.H. Brewer, E. Koster, A. Schenck, H. Schilling, D.L. Williams: Hyperfine Int. *8*, 671 (1981)
3.26 A.T. Fiory, K.G. Flynn, D.M. Parkin, W.J. Kossler, W.F. Lankford, C.E. Stronach: Phys. Rev. Lett. *40*, 968 (1978)
3.27 R.H. Heffner, J.A. Brown, R.L. Lutson, M. Leon, W.B. Gauster, O.N. Carlson, D.K. Rehbein, A.T. Fiory: Hyperfine Int. *6*, 237 (1979)
3.28 H. Schilling, M. Camani, F.N. Gygax, W. Rüegg, A. Schenck: Hpyerfine Int. *8*, 675 (1981)
3.29 W.F. Lankford, H.K. Birnbaum, A.T. Fiory, R.P. Minnich, K.G. Lynn, C.E. Stronach, L.H. Biemann, W.J. Kossler, J. Lindemuth: Hyperfine Int. *4*, 833 (1978)
3.30 O. Hartmann, E. Karlsson, L.O. Norlin, P. Pernestal, M. Borghini, T. Niinikoski, E. Walker: Hyperfine Int. *4*, 824 (1978)
3.31 H.K. Birnbaum, M. Camani, A.T. Fiory, F.N. Gygax, W.T. Kossler, W. Rüegg, A. Schenck, H. Schilling: Phys. Rev. B *17*, 4143 (1978)
3.32 M. Camani, F.N. Gygax, W. Rüegg, A. Schenck, H. Schilling: Jahresbericht E58, Schweizerisches Institut für Nuklearforschung (1977)
3.33 K.W. Kehr, G. Honig, D. Richter: Z. Phys. B *32*, 49 (1978)
3.34 T. McMullen, E. Zaremba: Phys. Rev. B *18*, 3026 (1978)
3.35 S. Fujii, Y Uemura: Solid State Comm. *26*, 761 (1978)
3.36 I.I. Gurevich, E.A. Meleshko, I.A. Muratova, B.A. Kikolsky, V.S. Roganov, V.I. Selivanov, B.V. Sokolov: Phys. Lett. *40A*, 143 (1972)
3.37 V.G. Grebinnik, I.I. Gurevich, V.A. Zhukov, A.P. Manych, E.A. Meleshko, I.A. Muratova, B.A. Nikolskii, V.I. Selivanov, V.A. Suetin: Sov. Phys. JETP *41*, 777 (1976)
3.38 L. Katz, M. Guinan, R.J. Borg: Phys. Rev. B *4*, 330 (1971)
3.39 O. Hartmann, E. Karlsson, L.O. Norlin, T.O. Niinikoski, K.W. Kehr, D. Richter, J.M. Welter, A. Yaounac, J. LeHericy: Phys. Rev. Lett. *44*, 337 (1980)
3.40 D. Richter: in *Nuclear and Electron Resonance Spectroscopies applied to Materials Science*, ed. by E.N. Kaufmann, G.K. Shenoy (North Holland, Yew York, Oxford, 1981)
3.41 M. Borghini, T.O. Niinikoski, J.C. Soulié, O. Hartmann, E. Karlsson, L.O. Norlin, K. Pernestal, K.W. Kehr, D. Richter, E. Walker: Phys. Rev. Lett. *40*, 1723 (1978)
3.42 V.G. Grebinnik, I.I. Gurevich, V.A. Zhukov, A.I. Klimov, V.N. Maiorov, A.D. Manych, E.V. Melnikov, B.A. Nikolskii, A.V. Pigorov, A.N. Ponomarev, V.I. Selivanov, V.A. Suetin: JETP Lett. *25*, 298 (1977)
3.43 T.O. Niinikoski, O. Hartmann, E. Karlsson, L.O. Norlin, K. Pernestal, K.W. Kehr, D. Richter, E. Walker, K. Schulze: Hyperfine Int. *6*, 229 (1979)
3.44 J.A. Brown, R.H. Heffner, M. Leon, D.M. Parkin, M.E. Schillaci, W.B. Gauster, A.T. Fiory, W.J. Kossler, H.K. Birnbaum, A.B. Denison, D.W. Cooke: Hyperfine Int. *6*, 233 (1979)
3.45 H. Metz, H. Orth, G. zu Putlitz, A. Seeger, H. Teichler, J. Vetter, W. Wahl, M. Wigand, K. Dorenburg, M. Gladisch, D. Herlach: Hyperfine Int. *6*, 271 (1979)
3.46 O. Hartmann, E. Karlsson, R. Wäppling, D. Richter, R. Hempelmann, K. Schulze, B. Patterson, E. Holzschuh, W. Kündig, S. Cox: Phys. Rev. B, *27*, 1943 (1983)
3.47 W.J. Kossler, A.T. Fiory, W.F. Lankford, L. Lindemuth, K.G. Lynn, S. Mahajan, R.P. Minnich, K.G. Petzinger, C.E. Stronach: Phys. Rev. Lett. *41*, 1558 (1978)
3.48 V.G. Grebinnik, I.I. Gurevich, A.Yu. Didyk, V.A. Zhukov, A.P. Manych, E.V. Melnikov, B.A. Nikolskii, V.S. Roganov, V.I. Selivanov, V.A. Suetin: JETP Lett. *27*, 30 (1978)
3.49 W.J. Kossler, A.T. Fiory, W.F. Lankford, K.G. Lynn, R.P. Minnich: Hyperfine Int. *6*, 295 (1979)
3.50 K. Dorenburg, M. Gladisch, D. Herlach, W. Mansel, H. Metz, H. Orth, G. zu Putlitz, A. Seeger, W. Wahl, M. Wigand: Z. Phys. B *31*, 165 (1978)
3.51 O. Hartmann, L.O. Norlin, K.W. Kehr, D. Richter, J.M. Welter, E. Karlsson, T.O. Niinikoski, A. Yaouanc: in *Recent Developments in Condensed Matter Physics*, Vol. 2, ed. by J.T. Devreese (Plenum Press, New York, 1981)

3.52 K.W. Kehr, D. Richter, J.M. Welter, O. Hartmann, L.O. Norlin, E. Karlsson, T.O. Niinikoski, J. Chappert, A. Yaouanc: Hyperfine Int. *8*, 681 (1981)
3.53 K.W. Kehr, D. Richter, G. Honig: Hyperfine Int. *6*, 279 (1979)
3.54 K. Petzinger: Hyperfine Int. *6*, 223 (1979)
3.55 W. Franck, A. Seeger: Appl. Phys. *3*, 61 (1974)
3.56 D. Herlach: in *Recent Developments in Condensed Matter Physics*, Vol. 1, ed. by J.T. Devreese (Plenum Press, New York, 1981)
3.57 K. Werner: private communication
3.58 H. Metz: Thesis, Universität Stuttgart, Germany (1980)
4.1 T. Graham: Phil. Trans. Roy. Soc. (London) *156*, 399 (1866)
4.2 For a recent review see, e.g., *Hydrogen in Metals I, II*, ed. by G. Alefeld, J. Völkl, Topics in Applied Physics, Vol. 28 (Springer, Berlin, Heidelberg, New York, 1978)
4.3 *Diffusion in Solids: Recent Developments*, ed. by A.S. Nowick, J.J. Burton (Academic Press, New York, 1975)
4.4 H.K. Birnbaum, C.A. Wert: Ber. Bunsenges. Phys. Chem. *76*, 806 (1972)
4.5 J. Völkl, G. Alefeld: in ref. 4.2, Vol. 1, pp. 321
4.6 J. Völkl, G. Wollenweber, K.H. Klatt, G. Alefeld: Z. Naturforsch. *26a*, 922 (1971)
4.7 M. Glugla: Diploma Thesis, Münster (1980)
4.8 H. Wipf, G. Alefeld: phys. stat. sol. (a) *23*, 175 (1974)
4.9 D. Richter, B. Alefeld, A. Heidemann, N. Wakabayashi: J. Phys. F *7*, 569 (1977)
4.10 W. Gissler, G. Alefeld, T. Springer: J. Phys. Chem. Sol. *31*, 2361 (1970)
4.11 D.G. Westlake, S.T. Ockers, D.W. Regan: J. Less. Comm. Metals *49*, 341 (1976)
4.12 G. Schaumann, J. Völkl, G. Alefeld: phys. stat. sol. *42*, 401 (1970)
4.13 N. Boes, H. Züchner: Z. Naturforsch. *31a*, 760 (1976)
4.14 J. Völkl, H.C. Bauer, U. Freudenberg, M. Kokkinidis, G. Lang, K.-A. Steinhauser, G. Alefeld: *Internal Friction and Ultrasonic Attenuation in Solids*, ICFUAS-6-485 (University of Tokyo Press, 1977)
4.15 F.M. Mazzolai, H. Züchner: Z. Phys. Chem. NF *124*, 59 (1981)
4.16 O.N. Salmon, D. Randall: USAEC Report KAPL-984 (1954)
4.17 H. Pfeiffer, H. Peisl: Phys. Lett. *60A*, 363 (1977)
4.18 A.M. Stoneham: J. Nucl. Mat. *69*, 109 (1978)
4.19 V. Lottner, A. Heim, K.W. Kehr, T. Springer: IAEA Report SM-219/27, Vienna (1978)
4.20 G. Bohmholdt, E. Wicke: Z. Phys. Chem. N.F. *56*, 133 (1967)
4.21 G. Sicking: Ber. Bunsenges. *76*, 790 (1972)
4.22 H. Teichler: Z. Phys. Chem. N.F. *114*, 155 (1979)
4.23 see, e.g., T. Springer: *Quasielastic Neutron Scattering for the Investigation of Diffusion Motions in Solids and Liquids*, in Springer Tracts in Modern Physics, Vol. 64 (Springer, Berlin, Heidelberg, New York, 1972)
4.24 L. van Hove: Phys. Rev. *95*, 249 (1954)
4.25 C.T. Chudley, R.J. Elliott: Proc. Phys. Soc. *77*, 353 (1961)
4.26 J.M. Rowe, J.H. Rush, H.G. Smith, M. Mostoller, H. Flotow: Phys. Rev. Lett. *33*, 1297 (1974)
4.27 G. Blaesser, J. Peretti: Proc. Int. Conf. Vacancies and Interstitials in Metals, Jül. Conf. 2, Vol. 2 (KFA Jülich 1968), p. 886
4.28 L.A. de Graaf, J.J. Rush, H.E. Flotow, J.M. Rowe: J. Chem. Phys. *56*, 4574 (1972)
4.29 J.M. Rowe, J.J. Rush, H.E. Flotow: Phys. Rev. B *9*, 5039 (1974)
4.30 N. Stump, W. Gissler, R. Rubin: phys. stat. sol. (b) *54*, 295 (1972)
4.31 N. Wakabayashi, G. Alefeld, K.W. Kehr, T. Springer: Solid State Comm. *15*, 503 (1974)
4.32 W. Gissler, N. Stump: Physica *65*, 109 (1973)
4.33 V. Lottner, J.W. Haus, A. Heim, K.W. Kehr: J. Phys. Chem. Sol. *40*, 557 (1979)
4.34 V. Lottner, A. Heim, T. Springer: Z. Phys. B *32*, 157 (1978)
4.35 D. Emin: Phys. Rev. Lett. *25*, 1751 (1970)
4.36 D. Emin: Phys. Rev. B *3*, 1321 (1971)

4.37 P. Kopfstadt, W.E. Wallace, L.J. Hyvönen: J. Am. Chem. Soc. *81*, 5015, 5019 (1959)
4.38 O.J. Kleppa, P. Dantzer, M.E. Melnichak: J. Chem. Phys. *61*, 4048 (1974)
4.39 G. Pfeiffer, H. Wipf: J. Phys. F *6*, 167 (1976)
4.40 W. Münzing, J. Völkl, H. Wipf, G. Alefeld: Scripta Metall. *8*, 1327 (1975)
4.41 C. Baker, H.K. Birnbaum: Acta Metall. *21*, 865 (1973)
4.42 P.E. Zapp, H.K. Birnbaum: Acta Metall. *28*, 1275, 1523 (1980)
4.43 A. Magerl, J.J. Rush, J.M. Rowe, D. Richter, H. Wipf: Phys. Rev. B *27*, 927 (1983)
4.44 D. Richter, J.J. Rush, J.M. Rowe: Phys. Rev. B, in print; see also D. Richter: J. Less Comm. Met. *89*, 293 (1983)
4.45 D. Richter: in ref. 2.12, pp. 245
4.46 D. Richter, K.W. Kehr, T. Springer: in *Proceedings of the Conference on Neutron Scattering*, Gatlinburg TN, ed. by R.M. Moon (Conf.-760601-PI, Vol. 1, 568 (1976))
4.47 K.W. Kehr, D. Richter, K. Schröder: IAEA Report SM-219, Vienna (1978) pp. 399
4.48 Zh. Qi, J. Völkl, H. Wipf: Scripta Metall. *16*, 859 (1982)
4.49 R.H. Swendsen, D. Richter, K.W. Kehr: unpublished
4.50 J. Völkl, G. Alefeld, Z. Phys. Chem. N.F. *114*, 123 (1979)
4.51 D. Richter, A. Kollmar: *Annex of the Annual Report of the ILL 1977*, Grenoble, France (1978)
4.52 T. Matsumoto: J. Phys. Soc. Jpn. *42*, 1583 (1977)
4.53 G.E. Murch: J. Nucl. Mat. *57*, 239 (1975)
4.54 G.E. Murch, R.J. Thorn: J. Phys. Chem. Sol. *38*, 789 (1977)
4.55 D.K. Ross, D.T. Wilson: IAEA Report SM-219/80, Vienna (1978)
4.56 K.W. Kehr, R. Kutner, K. Binder: Phys. Rev. B *23*, 4931 (1981)
4.57 A.D. LeClaire: in *Physical Chemistry*, Vol. 10, ed. by H. Eyring, D. Henderson, W. Jost (Academic Press, New York, 1970) pp. 261
4.58 D. Wolf: Phys. Rev. B *10*, 2710 (1974)
4.59 D. Wolf: Solid State Comm. *23*, 583 (1977)
4.60 A.D. LeClaire: Phys. Chem. (Solid State) *10*, 26 (1970)
4.61 I.S. Anderson, D.K. Ross, C.J. Carlile: IAEA Report SM-219/40, Vienna (1978)
4.62 I.S. Anderson, C.J. Carlile, D.K. Ross, D.L.T. Wilson: Z. Phys. Chem. N.F. *115*, 165 (1979)
4.63 see, e.g., R.F. Karlicek Jr., I.J. Lowe: Solid State Comm. *31*, 163 (1979)
4.64 B. Alefeld: Kerntechnik *14*, 15 (1972)
4.65 E. Lebsanft, D. Richter, J. Töpler: J. Phys. F *9*, 1057 (1979)
4.66 J. Töpler, E. Lebsanft, R. Schätzler: J. Phys. F *8*, L25 (1978)
4.67 E. Lebsanft, D. Richter, J. Töpler: Z. Phys. Chem. N.F. *116*, 175 (1979)
4.68 D. Richter, R. Hempelmann, L.A. Vinhas: J. Less Comm. Met. *88*, 353 (1983)
4.69 R. Hempelmann, D. Richter, A. Heidemann: J. Less Comm. Met. *88*, 343 (1983)
4.70 R. Hempelmann, D. Richter, R. Pugliesi, L.A. Vinhas: J. Phys. F *13*, 59 (1983)
4.71 T. Springer: in ref. 2.2, Vol. 1, pp. 75
4.72 J.M. Rowe, J.J. Rush, H.G. Smith, M. Mostoller, H.E. Flotow: Phys. Rev. Lett. *33*, 1297 (1974)
4.73 N. Stump, G. Alefeld, D. Tochetti: Solid State Comm. *19*, 805 (1976)
4.74 D. Richter, S.M. Shapiro: Phys. Rev. B *22*, 599 (1980)
4.75 J. Eckert, J.A. Goldstone, D. Tonks, D. Richter: Phys. Rev. B *27*, 1980 (1983)
4.76 G. Herzberg: in *Molecular Spectra and Molecular Structure I* (Van Nostrand, New York, 1950) p. 92ff
4.77 B.N. Ganguly: Z. Phys. *265*, 433 (1973) and Phys. Rev. B *14*, 3848 (1976)
4.78 R.J. Müller, C.B. Satterthwaite: Phys. Rev. Lett. *34*, 144 (1975)
4.79 J.J. Rush, A. Magerl, J.M. Rowe, J.M. Harris, J.L. Provo: Phys. Rev. B *24*, 4902 (1981)
4.80 R. Hempelmann, D. Richter, A. Kollmar: Z. Phys. B *44*, 159 (1981)
4.81 K.W. Kehr: in ref. 4.2, Vol. 1, pp. 197
4.82 V. Lottner, H.R. Schober, W.J. Fitzgerald: Phys. Rev. Lett. *42*, 1162 (1979)
4.83 H.R. Schober, V. Lottner: Z. Phys. Chem. N.F. *114*, 203 (1979)
4.84 A. Magerl, B. Berre, G. Alefeld: phys. stat. sol. (a) *36*, 161 (1976)

4.85 M.R. Chowdhury: J. Phys. F *4*, 1657 (1974)
4.86 H. Sugimoto, Y. Fukai: Phys. Rev. B *22*, 670 (1980)
4.87 R.F. Mattas, H.K. Birnbaum: Acta Metall. *23*, 973 (1975)
4.88 P. Schiller, N. Nijman: phys. stat. sol. *31*, K77 (1975)
4.89 G. Canelli, L. Verdini: Richerca Sci. *36*, 98 (1966)
4.90 C.G. Chen, H.K. Birnbaum: phys. stat. sol. (a) *36*, 687 (1976)
4.91 G. Canelli, R. Cantelli: Proceedings of the International Symposium on Metal-Hydrogen System, 13-15 April 1971, Miami Beach, USA
4.92 G.J. Sellers, A.C. Anderson, H.K. Birnbaum: Phys. Rev. B *10*, 2771 (1974)
4.93 C. Morkel, H. Wipf, K. Neumayer: Phys. Rev. Lett. *40*, 947 (1978)
4.94 M. Locatelli, K. Neumeier, H. Wipf: J. Physique *39*, C6-995 (1978)
4.95 D.B. Poker, G.G. Setser, A.V. Granato, H.K. Birnbaum: Z. Phys. Chem. N.F. *116*, 39 (1979)
4.96 H. Wipf, A. Magerl, S.M. Shapiro, S.K. Satija, W. Thomlinson: Phys. Rev. Lett. *46*, 947 (1981)
4.97 D. Richter, H. Wipf: to be published
4.98 A.S. Nowick, W.R. Heller: Adv. Phys. *12*, 251 (1963) and *14*, 101 (1965)
4.99 A.S. Nowick: Adv. Phys. *16*, 1 (1967)
4.100 H.K. Birnbaum, C.P. Flynn: Phys. Rev. Lett. *37*, 25 (1976)
4.101 D. Richter, G. Alefeld, H. Wipf, A. Magerl: *Annex of the Annual Report ILL 1978*, Grenoble (1979)
4.102 Y. Imry: in *Tunneling Phenomena in Solids*, ed. by E. Burnstein, S. Lindquist (Plenum, New York, 1969)
4.103 see, e.g., W. Marshall, S.W. Lovesey: *Theory of Thermal Neutron Scattering*, (Clarendon Press, Oxford, 1971)
4.104 A.M. Stoneham: Rev. Mod. Phys. *41*, 82 (1969)
4.105 Y. Fukai, H. Sugimoto: Proceedings JIMIS-2 on Hydrogen in Metals (Tokyo, 1980)
4.106 H.D. Carstanjen: phys. stat. sol. (a) *59*, 11 (1980)
5.1 see, e.g., A.S. Davidov: *Theory of Molecular Excitons* (Plenum, New York, 1971) and
5.2 R.W. Munn, R. Silbey: J. Chem. Phys. *68*, 2439 (1978)
5.3 see, e.g., N. Giordano, W. Gilson, D.E. Prober: Phys. Rev. Lett. *43*, 725 (1979) or
5.4 D.J. Thouless: Phys. Rev. Lett. *39*, 1167 (1977)
5.5 see, e.g., *Positrons in Solids*, ed. by P. Hautojärvi, in Topics in Current Physics, Vol. 12 (Springer, Berlin, Heidelberg, New York, 1979)
5.6 see, e.g., W. Press: *Single Particle Rotations in Molecular Crystals*, in Springer Tracts in Modern Physics, Vol. 92 (Springer, Berlin, Heidelberg, New York, 1981)

Combined Subject Index

Springer Tracts in Modern Physics

Editor: **G. Höhler** · Associate Editor: **E. A. Niekisch**

Volume 93

B. Dorner

Coherent Inelastic Neutron Scattering in Lattice Dynamics

1982. 47 figures. VIII, 96 pages
ISBN 3-540-11049-6

Contents: Introduction. – Experimental Technique with There-Axis Spectrometers. – The Scattering Function and Symmetry Operations in the Crystal. – Lattice Dynamical Models. – Analysis of Phonon Intensities. – Analysis of Phonon Line Shapes. – Final Remarks. – References. – Subject Index.

Volume 65

1972. 81 figures. III, 145 pages
ISBN 3-540-05876-1

Contents: *H. Theissen:* Spectroscopy of Light Nuclei by Low Energy (< 70MeV) Inelastic Electron Scattering. – *H. Arenhövel, H. J. Weber:* Nuclear Isobar Configurations. – *K. Heinloth:* Experiments on Electroproduction in High Energy Physics.

Volume 64

T. Springer

Quasielastic Neutron Scattering for the Investigation of Diffusive Motions in Solids and Liquids

1972. 36 figures. II, 100 pages
ISBN 3-540-05808-7

Contents: Scattering Theory. – Methodical and Experimental Aspects. – Monoatomic Liquids with Continuous Diffusion. – Jump Diffusion in Liquids. – Diffusion of Hydrogen in Metals. – Rotational Diffusion in Molecular Solids. – Molecular Liquids. – Polymers and other Complicated Systems. – Effects of Coherent Scattering. – Quasielastic Scattering and other Methods.

Springer-Verlag
Berlin
Heidelberg
New York
Tokyo